Sustainability in Biodegradation

Sustainability in Biodegradation

Edited by **William Chang**

New York

Published by Callisto Reference,
106 Park Avenue, Suite 200,
New York, NY 10016, USA
www.callistoreference.com

Sustainability in Biodegradation
Edited by William Chang

International Standard Book Number: 978-1-63239-576-4 (Hardback)

Printed in the United States of America.

Contents

Preface

Every book is a source of knowledge and this one is no exception. The idea that led to the conceptualization of this book was the fact that the world is advancing rapidly; which makes it crucial to document the progress in every field. I am aware that a lot of data is already available, yet, there is a lot more to learn. Hence, I accepted the responsibility of editing this book and contributing my knowledge to the community.

This book provides an array of various research works discussing different technologies that have been used for the escalation of biodegradation process. The book deals with various factors and aspects of biodegradation. These include biodegradation and sustainability.

While editing this book, I had multiple visions for it. Then I finally narrowed down to make every chapter a sole standing text explaining a particular topic, so that they can be used independently. However, the umbrella subject sinews them into a common theme. This makes the book a unique platform of knowledge.

I would like to give the major credit of this book to the experts from every corner of the world, who took the time to share their expertise with us. Also, I owe the completion of this book to the never-ending support of my family, who supported me throughout the project.

Editor

Biodegradation and Sustainability

Methods for Separation, Recycling and Reuse of Biodegradation Products

Ganapati D. Yadav and Jyoti B. Sontakke

Additional information is available at the end of the chapter

1. Introduction

Thousands of chemicals and materials with varied properties and functionalities are manufactured and used for commercial and day-to-day applications, whose ultimate fate in the environment may not be known. During their manufacture and use, these substances are often discharged into the environment through different routes in air, water and land. Creation of tremendous quantities of solid waste of all kind and its effective disposal has posed innumerable problems that need technological breakthroughs. Many of these substances degrade slowly and exert toxic effects on plants and animals, thus causing large scale environmental degradation [1, 2]. Pollution by abandoned plastic articles is also a matter of great concern [3]. Industrial wastewaters associated with the manufacture of organic chemicals are voluminous and characteristically have concentrations ranging from a few ppm to a thousands of ppm. Biodegradation of such dissolved pollutants is an area of immense interest to various sectors. Emission of volatile organic compounds (VOCs) from various sources has detrimental effects on quality of air we breathe and on environmental phenomena. Biodegradation, either aerobic or anaerobic, can be an approach to cleave big molecules through a series of steps in to smaller molecules from a mosaic of chemicals and materials and some of them can be valorized as pollution abatement strategy and source of energy through biogas generation [2]. Biogas can be produced from nearly all kind of biomass, among which the primary agricultural sectors and various organic waste streams can be properly tapped as renewable source of energy. Untreated or poorly managed animal manure is a major source of air and water pollution. Nutrient leaching, mainly nitrogen and phosphorous, ammonia evaporation and pathogen contamination are some of the foremost threats [3]. A conservative estimate is provided by Steinfeld et al. [4] that the animal production sector is responsible for 18% of the overall green house gas emissions, measured in CO_2 equivalent and for 37% of the anthropogenic methane, which has 23 times the global warming potential of CO_2. Furthermore, 65% of anthropogenic nitrous oxide and 64% of anthropogenic

ammonia emission originate from the worldwide animal production sector. Biogas production from anaerobic digestion of animal manure and slurries can be harnessed to alleviate
greenhouse gas emissions in particularly ammonia and methane [5].

Plastics are bane and benefactor simultaneously. Over 230 million tons of plastic are produced
annually. Plastics are used in all walks of life and provide improved insulation, lighter
packaging, are found in cars, aeroplanes, railways, phones, computers, medical devices, etc.
but appropriate disposal is often not properly addressed. On one hand, plastic waste and
disposal is a hotly debated issue globally whereas on the other, it can contribute to reduce the
carbon footprint. Many leading European countries recover more than 80% of their used
plastics, by adopting an integrated waste and resource management strategy to address each
waste stream with the best options [6]. Plastic sorting and separation, recycling, depolymeri-
sation, cracking, and production of fuel are some of the strategies used to abate plastic
pollution. Development of biopolymers is pursued vigorously. Biodegradation of plastics by
microorganisms and enzymes appears to be the most effective process. When plastics are used
as substrates for microorganisms, evaluation of their biodegradability should not only be based
on their chemical structure, but also on their physical properties such as melting point, glass
transition temperature, crystallinity, storage modulus, etc. [7-11].

This chapter has covered the mechanisms of biodegradation, biodegradation of a variety of
industrial chemicals, plastics and other biomass, advances in anaerobic digestion technologies
and biogas generation, plastic processing, biopolymer synthesis and degradation. Synthesis
of biopolymers is covered. The scope for treating municipal organic solid waste, manure and
polymers to generate biogas as a renewable energy option, and also as a pollution abatement
strategy is discussed including technological aspects. The synthesis of biohydrogen, bioetha-
nol, biobutanol and other biotransformation leading to valuable chemicals, which also involve
breaking down of larger molecules, plastics and biomaterials are not addressed [7,10].
Biorefinery is a concept which is akin to petrorefinery, wherein biomass is converted into useful
platform chemicals through extraction, controlled pyrolysis, fermentation, enzyme and
chemical catalysis [12].

2. Mechanisms of biodegradation

Cellulose, lignocellulose and lignin are major sources of plant biomass and are polymeric
substances; therefore, their recycling is indispensable for the carbon cycle [13]. Each of these
polymer is degraded by a variety of microorganisms which produce scores of enzymes that
work in tandem. The diversity of cellulosic and lignocellulosic substrates has contributed to
the difficulties found in enzymatic treatment. Fungi are the best-known microorganisms
capable of degrading these three polymers. Because the substrates are insoluble, both bacterial
and fungal degradation occur exo-cellularly, either in association with the outer cell envelope
layer or extra-cellularly. Microorganisms have two types of extracellular enzymatic systems,
namely, the hydrolytic system, which produces hydrolases and is responsible for cellulose and
hemicellulose degradation; and a unique oxidative and extracellular ligninolytic system,

which depolymerizes lignin [13]. The man-made chemicals and materials are comprised of different entities and functional groups which need to be degraded effectively by microorganisms and no single microorganism is obviously capable of doing it [1,14].

Growth and co-metabolism are the two mechanisms of biodegradation. In the case of growth, organic substance is used as the sole source of carbon and energy, which leads to complete degradation (mineralization). Archaebacteria, prokaryotes and eukaryotes (like fungi, algae, yeasts, protozoa) play dominant role in mineralization [7]. On the contrary, co-metabolism encompasses the metabolism of an organic compound in the presence of a growth substrate which is used as the primary carbon and energy source. Thus, biodegradation processes and their rates differ greatly depending on the type of substrate and conditions such as temperature, pH, and aqueous phase solubility, but frequently the major final products of the degradation are carbon dioxide and methane [1,7,10].

2.1. Growth-associated degradation of aliphatic compounds

Growth-associated degradation produces CO_2, H_2O, and cell biomass. The cells act as the complex biocatalysts of degradation. Further, cell biomass may be mineralized after exhaustion of the degradable pollutants in a contaminated site. Bulk chemicals like aromatic hydrocarbons such as benzene, toluene, ethylbenzene, xylenes, and naphthalene are widely used as fuels, industrial solvents and feedstock for petrochemical industry. Phenols and chlorophenols are another class of chemicals, employed in a variety of industries. Since all micro-organisms make aromatic compounds such as aromatic amino acids, phenols, or quinines, in large amounts, many microorganisms have evolved catabolic pathways to degrade aromatic compounds. In general, man-made organic chemicals (xenobiotics) can be degraded by microorganisms, when the respective molecules are similar to natural compounds [7,10].

In general, benzene, condensed ring and related compounds are characterized by a higher thermodynamic stability than aliphatic compounds. Benzene oxidation begins with hydroxylation catalyzed by a dioxygenase leading to a diol (Scheme 1) which is then converted to catechol by a dehydrogenase.

Hydroxylation and dehydrogenation are also common in degradation routes of other aromatic hydrocarbons. The introduction of a substituent group onto the benzene ring renders alternative mechanisms possible to attack side chains or to oxidize the aromatic ring. Many aromatic substrates are degraded by a limited number of reactions such as hydroxylation, oxygenolytic ring cleavage, isomerization, and hydrolysis. The inducible nature of the enzymes and their substrate specificity enable bacteria such as *pseudomonads* and *rhodococci* with a high degradation activity, to acclimatize their metabolism to the effective utilization of substrate mixtures in polluted soils and also to grow at a high rate [10,15].

2.2. Co-metabolic degradation of organo-pollutants

Co-metabolism is a common phenomenon of microbial activities and the basis of biotransformation used in biotechnology to convert molecules in to useful modified forms. Microorganisms growing on a particular substrate also oxidize a second substrate. The co-substrate is not

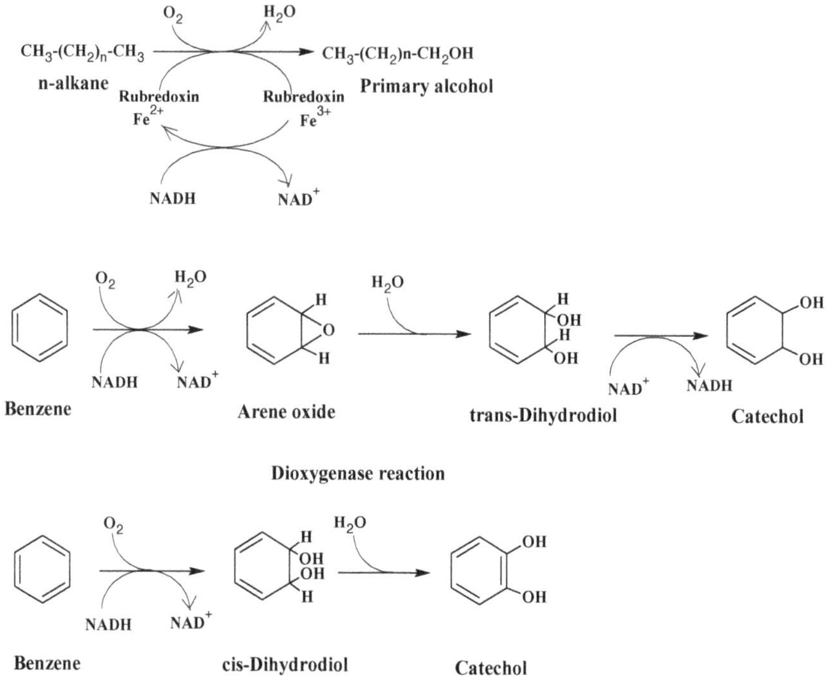

Scheme 1. Monooxygenase and dioxygenase reactions: In this mechanism, monooxygenase initially incorporates one O atom from O_2 into the xenobiotic substrate whereas the other is reduced to H_2O. On the contrary, dioxygenase incorporates both atoms into the substrate [15].

incorporated, but the product may be available as substrate for other organisms of a mixed culture. The rudiments of co-metabolic transformation are the enzymes of the growing cells and the synthesis of cofactors necessary for enzymatic reactions; for instance, of hydrogen donors (reducing equivalents, NADH) for oxygenases. Several aromatic substrates can be converted enzymatically to natural intermediates of degradation such as catechol and protecatechuate (Scheme 2) [15].

Co-metabolism of chloroaromatics is a general activity of bacteria in mixtures of industrial pollutants. The co-metabolic transformation of 2-chlorophenol leads to dead-end metabolites such as 3-chlorocatechol, which may be auto-oxidized or polymerized in soil to humic-like structures. Irreversible binding of dead end metabolites may fulfill the function of detoxification. The accumulation of dead-end products within microbes under selection pressure is the source for the evolution of new catabolic traits. Thus, recalcitrance of organic pollutants increases with increasing halogenation. Substitution of halogen as well as nitro and sulfo groups at the aromatic ring is accomplished by an increasing electrophilicity of the molecule. These compounds resist the electrophilic attack by oxygenases of aerobic bacteria. Compounds that persist under oxic condition are polychlorinated biphenyls (PCBs), chlorinated dioxins

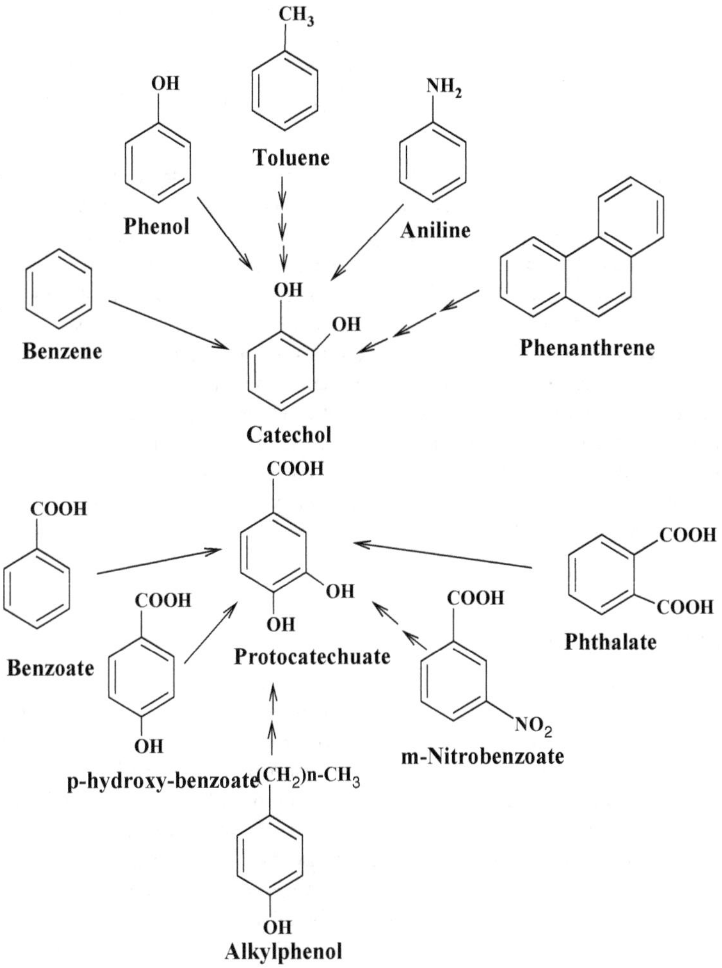

Scheme 2. Degradation of aromatic natural and xenobiotic compounds into two central intermediates, catechol and protocatechuate [15].

and some pesticides like DDT. To overcome the relatively high persistence of halogenated xenobiotics, reductive attack of anaerobic bacteria is of great value. Reductive dehalogenation achieved by anaerobic bacteria is either a gratuitous reaction or a new type of anaerobic respiration. The process reduces the degree of chlorination and, therefore, makes the product more accessible to mineralization by aerobic bacteria [7,15].

Reductive dehalogenation which is the first step of degradation of PCBs requires anaerobic conditions wherein organic substrates act as electron donors. PCBs accept electrons to allow the anaerobic bacteria to transfer electrons to these compounds. Anaerobic bacteria capable of catalyzing reductive dehalogenation seem to be relatively omnipresent in nature. Most dechlorinating cultures are a mixed consortia. Anaerobic dechlorination is always incomplete and the products are di- and monochlorinated biphenyls. These products can be metabolized further by aerobic microorganisms [2,7,15].

The rates of biodegradability of particular substrate is mainly related to accessibility of the substrate for enzymes and can be enhanced by several means as reviewed by van Lier et al. [16] such as (a) mechanical methods: the disintegration and grinding of solid particles present in sludge: releases cell compounds and creates new surface where biodegradation take place, (b) ultrasonic disintegration, (c) chemical methods: the destruction of complex organic compounds by means of strong, mineral acids or alkalis, (d) thermal pretreatment: thermal hydrolysis is able to split and decompose a significant part of the sludge solid fraction into soluble and less complex molecules, (e) enzymatic and microbial pre-treatment: a very promising method for the future for some specific substrates (e.g. cellulose, lignin etc.),(f) stimulation of anaerobic micro-organisms: some organic compounds (e.g. amino acids, cofactors, cell content) act as a stimulating agent in bacteria growth and methane production. Most of the above methods occur at the pre-methanation step and result in a better supply of methanogenic bacteria by suitable substrates.

3. Aerobic biodegradation

Many microorganisms grow under aerobic conditions. The so-called cellular respiration process (CSP) begins with aerobes which employ oxygen to oxidize substrates such as sugars and fats to derive energy. Before the onset of CSP, glucose molecules are degraded into smaller molecules in the cytoplasm of the aerobes. The smaller molecules then enter a mitochondrion, where aerobic respiration takes place. Oxygen is used to break down small entities into water and carbon dioxide, accompanied by release of energy. Aerobic degradation does not produce foul gases, unlike anaerobic process. The aerobic process leads to a more complete digestion of solid waste reducing build-up by more than 50% in most cases [1, 2, 7]. The major enzymatic reactions of aerobic biodegradation are oxidations catalyzed by oxygenases and peroxidases. Oxygenases are oxido-reductases that incorporate oxygen into the substrate as given in Scheme 1. Degradative organisms need oxygen at two metabolic sites, namely, at the initial attack of the substrate and at the end of the respiratory chain. Higher fungi possess a unique oxidative system for the degradation of lignin based on extracellular ligninolytic peroxidases and laccases [13]. This enzymatic system is important for the co-metabolic degradation of persistent organic pollutants. The predominant bacteria of polluted soils belong to a spectrum of genera and species (Table 1) [15].

The most important classes of organic pollutants in the environment are mineral oil constituents and halogenated petrochemicals, for the biodegradation of which the capacities of aerobic microorganisms are of great consequence. The most rapid and complete degradation of the majority of pollutants is brought about under aerobic conditions and these include petroleum hydrocarbons, chlorinated aliphatics, benzene, toluene, phenol, naphthalene, fluorine, pyrene, chloroanilines, pentachlorophenol and dichlorobenzenes. Many cultures of bacteria grow on these chemicals and are capable of producing enzymes which degrade them into non-toxic species. [7,15].

Gram negative bacteria	Gram positive bacteria
Pseudomonas species	*Nocardia* species
Xanthomonas species	*Mycobacteria* species
Alcaligenes species	*Corynebacterium* species
Flavobacterium species	*Arthobacter* species
Cytophaga group	*Bacillus* species

Table 1. Predominant bacteria in soil samples polluted with aliphatic and aromatic hydrocarbons, polycyclic aromatic hydrocarbons, and chlorinated compounds [15]

There are several essential attributes of aerobic microorganisms degrading organic pollutants amongst which metalobic processes top the list. The chemicals must be accessible to the degrading organisms. For example, hydrocarbons are immiscible in water and their degradation requires the production of biosurfactants in order to have effective biodegradation [14]. The initial intracellular attack of organic pollutants is an oxidative process and therefore, the activation and incorporation of oxygen is the main enzymatic reaction catalyzed by oxygenases and peroxidases. Peripheral degradation pathways convert organic pollutants step by step into intermediates of the central intermediary metabolism, such as the tricarboxylic acid cycle. Biosynthesis of cell biomass from the central precursor metabolites (acetyl-CoA, succinate, pyruvate) is required [14,15]. Sugars needed for various biosyntheses and growth must be synthesized by gluconeogenesis. The predominant degraders of organo-pollutants in the oxic zone of contaminated areas are chemo-organotropic species that are able to use a large number of natural and xenobiotic compounds as carbon sources and electron donors for the generation of energy. Although many bacteria are able to metabolize organic pollutants, a single bacterium does not possess the enzymatic capability to degrade all or even most of the organic pollutants from a heterogeneous mixture originating from particular industries. Thus, mixed microbial communities have the most powerful biodegradative potential. The genetic information of more than one organism is necessary to develop a system which could be used on industrial scale to degrade the complex mixtures of organic compounds present in contaminated areas. The genetic potential and certain environmental factors such as temperature, pH, and available nitrogen and phosphorus sources govern the rate and the extent of degradation [14].

4. Anaerobic biodegradation

Among biological treatments, anaerobic digestion is frequently the most economical process, due to the high energy recovery linked to the process and its limited environmental impact.

Anaerobic biodegradation results when the anaerobic microbes are predominant over the aerobic microbes. Here oxygen does not serve as the final electron acceptor or reactant. Manganese and iron ions, and substances like sulfur, sulfate, nitrate, carbon dioxide, some organic intermediates and pollutants are reduced by electrons originating from oxidation of organic compounds [7]. The common example of anaerobic process is the biodegradable waste in landfill. Paper and other materials degrade more slowly over longer periods of time. Biogas, coming from anaerobic digestion, mainly consists of methane and can be collected efficiently and used for eco-friendly power generation as has been demonstrated on larger scale [3, 16]. Anaerobic digestion is widely used, as part of an integrated waste management system, to treat wastewater sludge and biodegradable waste because it provides volume and mass reduction of the input material. It reduces the emission of landfill gas into the atmosphere [17-20]. Anaerobic digestion is a renewable energy source because the process produces methane and CO_2-rich biogas suitable for energy production helping to replace fossil fuel requirement. Also, the nutrient-rich solids left after digestion can be used as fertilizer [16,21].

There are four major biological and chemical steps of anaerobic digestion: hydrolysis, acido-genesis, acetogenesis, and methanogenesis [17,18]. The mechanism commences with bacterial hydrolysis of the organic matter to break down insoluble organic polymers such as carbohydrates and make them available for other bacteria. Acetogenic bacteria convert the sugars and amino acids into carbon dioxide, hydrogen, ammonia, and organic acid. Methanogens then ultimately transform these products in to methane and carbon dioxide [19].

4.1. Advances in anaerobic digestion technologies

Thermophilic anaerobic digestion of manure [20] and assessment of biodegradability of macropollutants [21] have demonstrated the prowess of anaerobic digestion which is now a general method used to stabilize municipal wastewater treatment residuals [22,23]. The so-called phased or staged anaerobic digestion is a recent technology for digestion facilities which include four different configurations of reactors: staged mesophilic digestion, temperature-phased digestion, acid/gas phased digestion, and staged thermophilic digestion [24]. Phased or staged configurations are multiple reactor digestion systems. Phased anaerobic digestion is defined as a digestion system having two or more tanks, each with exclusive operating conditions that support unique biomass populations, which may be acid-forming, methane-forming, thermophilic, or mesophilic organism populations. Effective digestion is achieved by manipulating operational parameters such as solids retention time (SRT) and temperature. Temperature phased digestion system is found better than the other systems during each study phase by having higher volatile solids reduction (VSR), higher methane production, and lower residual biological activity [24,25].

On industrial scale, anaerobic digestion of solid waste is considered as a mature technology [16,26]. Around 60% of the plants are reported in Europe to operate at the mesophilic range (40% thermophilic) with continued increase in capacity over the years in most European countries. Yields from the biomethanization process are very much dependent on operating conditions and the kind of substrate used. Digestion of grey wastes or residual refuse after source separation, has caught attention of industry and some of the solutions considered are landfilling or incineration [23]. However, anaerobic digestion is a better option since it gives number of advantages such as greater flexibility, the possibility of additional material recovery (up to 25%), and a more efficient and ecological energy recovery. In this case the low-calorific organic fraction is digested, the high-calorific fraction is treated thermally and the non-energy fractions can be recovered and reused. It is predicted that this residual refuse will be treated by anaerobic digestion [16, 23].

A very high growth potential is expected for the anaerobic digestion of organic fraction of municipal solid waste (OFMSW). Around 50% of MSW is landfilled, with a content of around 30% of organic fraction (without considering paper and cardboard). The growth potential for this technology is very important to reduce greenhouse gases emission as agreed at the Kyoto Summit [23]. Further, the consolidation of anaerobic digestion as a mainstream technology for the OFMSW should occur since the digested residue can be considered quite stable organic matter with a very slow turnover of several decades given adequate soil conditions. Thus, the natural imbalance in CO_2 can be adjusted by restoring or creating organic rich soil. The removal of CO_2 constitutes an extra benefit that would place anaerobic digestion as one of the most relevant technologies in this field. The degradation of chlorinated compounds need to be examined in greater depth, as anaerobic treatment offers high potential in this area [28].

Several novel reactors with high mass transfer rates, such as fluidized bed reactors, expanded granular sludge bed (EGSB) reactors [29-32], and membrane bioreactors [33] with different configurations have been used, in which hydraulic retention times (HRT) are uncoupled from the solids retention time (SRT) to make anaerobic technology economical alternative for conventional wastewater treatment systems. The upflow anaerobic sludge blanket (UASB) reactors [30] and/or related systems are mostly applied, wherein spontaneous formation of granular conglomerates of the anaerobic organisms occurs, leading to anaerobic sludge with an extremely low sludge volume index and optimal settling properties [21]. Besides, several large scale biogas plants have been built which combine waste from agriculture, industry and households and produce both biogas and a liquid fertiliser which is re-circulated back on agri-land. The combination of anaerobic digestion with other biological or physico-chemical processes has led to the development of optimised processes for the combined removal of organic matter, sulphur and nutrients. In conjunction with anaerobic digestion which removes mainly carbon, other processes are used to remove nitrogen and phosphorus (with oxic phase), which mainly use micro-organisms and also physico-chemical processes. For the treatment of municipal wastewater, the ANANOX process [34] takes advantage of sulphate reduction to sulphide to provide an electron donor for the denitrification process [35-37]. The integration of the nitrogen cycle in anaerobic digestion could be maximised with the application of the ANAMMOX process that makes use of particular micro-organisms that are able to oxidise ammonium to N_2 gas with nitrite as electron acceptor [38,39].

5. Biodegradation of industrial organic pollutants

Knowledge of fate of chemicals discharged in the environment, the life cycle analysis and the mechanisms by which they degrade are of great importance in designing biodegradation systems since many of the industrial chemicals are toxic, recalcitrant and bioaccumulating in organisms [40-42].

5.1. Volatile Organic Compounds (VOCs)

There are two classes of VOCs that are responsible for a large number of land and groundwater contamination: (i) petroleum hydrocarbons (PHCs) such as gasoline, diesel, and jet fuel, and (ii) chlorinated hydrocarbon (CHC) solvents such as the dry cleaning agents such as tetrachloroethylene, perchloroethylene (PCE) and the degreasing solvents such as trichloroethylene (TCE), 1,1,1-trichloroethane (TCA), and PCE.

PHCs biodegrade readily under aerobic medium, whereas CHCs characteristically biodegrade much more slowly and under anaerobic conditions [43]. Because PHC biodegradation is relatively rapid when oxygen is present, aerobic biodegradation can usually limit the concentration and subsurface migration of petroleum vapours in unsaturated soils. Further, CHC biodegradation can produce toxic moieties, such as dichloroethylene and vinyl chloride, while petroleum degradation usually produces carbon dioxide, water, and sometimes methane or other simple hydrocarbons. A second primary difference is density of pollutant. PHC liquids are lighter than water and immiscible. PHCs can float on the groundwater surface (water table), whereas chlorinated solvents being heavier than water sink through the groundwater column to the bottom of the aquifer. These major differences in biodegradability and density lead to very different subsurface behaviour that often reduces the potential for human exposure.

5.1.1. Petroleum Hydrocarbons (PHCs)

It is known that microorganisms capable of aerobically degrading PHCs are present in nearly all subsurface soil environments [44-49]. Effective aerobic biodegradation of PHCs hinges on the soil having adequate oxygen and water content to provide a habitat for sufficient populations of active microorganisms. If oxygen is present, these organisms will generally consume available PHCs. Furthermore, aerobic biodegradation of petroleum compounds can occur relatively quickly, with degradation half lives as short as hours or days under some conditions [50]. Some petroleum compounds can also biodegrade under anaerobic conditions; however, above the water table, where oxygen is usually available in the soil zone, this process is insignificant and often much slower than aerobic biodegradation. Aerobic biodegradation consumes oxygen and generates carbon dioxide and water. This leads to a characteristic vertical concentration profile in the unsaturated zone in which oxygen concentrations decrease with depth and VOCs including PHCs and methane from anaerobic biodegradation and carbon dioxide concentrations increase with depth [51,52].

5.1.2. Chlorinated Hydrocarbon (CHC) Solvents

Chlorinated solvents such as tetrachloroethylene (TCE), 1,1,2,2-tetrachloroethane, carbon tetrachloride, and chloroform are released as waste products by spills, land-filling, and discharge to sewers during manufacture and their use as solvents in a variety of cleaning processes or as vehicles for solid slurries. TCE is a major pollutant of the industry. It is biodegraded under anaerobic conditions through hydrogenolysis that sequentially produces isomers of 1,2-dichloroethylene (1,2-DCE), vinyl chloride (VC), and ethylene. Some labs have also reported ethane [53,54], methane [55], and carbon dioxide [56] as degradation products.

In addition to anaerobic degradation through reductive dechlorination (hydrogenolysis), TCE and other chlorinated VOCs can be susceptible to co-metabolic oxidation by aerobic microorganisms that have oxygenases with broad substrate specificity. Methanotrophs are microorganisms that primarily oxidize methane for energy and growth using methane monooxygenase (MMO) enzymes and are a group of aerobic bacteria transform TCE through co-metabolic oxidation [57-59]. In contrast to reductive dechlorination, where the degradation rate generally decreases as the degree of chlorination of the aliphatic hydrocarbon decreases, the less-chlorinated VOCs such as 1,2-DCE and VC are more straightforwardly and quickly degraded through aerobic oxidation reactions than the higher chlorinated compounds such as TCE [60]. Methane-oxidizing bacteria are known to convert TCE to its epoxide, which then breaks down immediately in water to form dichloroacetic acid, glyoxylic acid, or one-carbon compounds such as formate or CO. The two carbon acids accumulate in the water phase, while formate and CO are further oxidized by methanotrophic bacteria to CO_2. Hence, coupling of anaerobic and aerobic degradation processes has been recommended as the best possible bioremediation method for chlorinated VOCs such as TCE [60-62].

5.2. Quinoline

Quinoline occurs commonly in coal tar, oil shale, and petroleum, and is used as an intermediate and solvent in many industries [63,64]. Due to its toxicity and repulsive odor, quinoline-containing waste is detrimental to human health and environmental quality. The study of quinoline- degrading bacteria not only helps to reveal the metabolic mechanism of quinoline, but also benefits the bio-treatment of quinoline-containing wastewater. Although different genera of bacteria may produce different intermediates, almost all of them transform quinoline into 2-hydroxyquinoline in the first step [63, 65]. A quinoline-degrading bacteria strain, *Pseudomonas* sp. BW003, was isolated from the activated sludge in a coking wastewater treatment plant. *Pseudomonas* strains degrade quinoline via the 2-hydroxyquinoline and 2,8-hydroquinoline pathway, and then transform 2,8-hydroquinoline into 8-hydrocumarin, which is then transformed into 2,3-dihydroxyphenylpropionic acid, and finally to CO_2 and H_2O (Scheme 3) [66-69]. Quinoline-N is transformed into ammonia-N, as reported in few genera of bacteria. Thus, quinoline pollution can be eliminated by applying such degrading bacteria in the treatment with bio-augmentation [70-72].

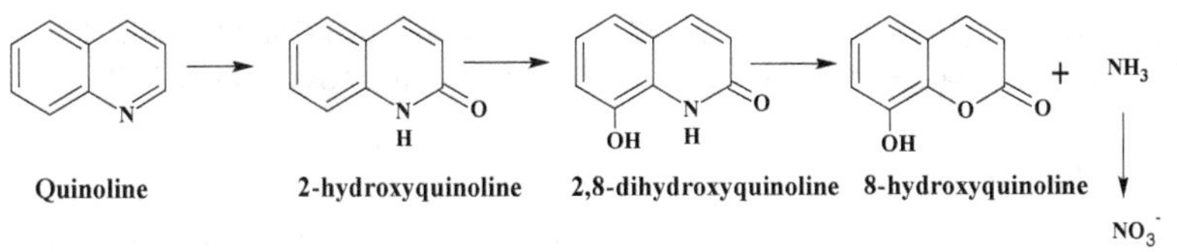

Scheme 3. Degradation products of quinoline [63]

5.3. Phenols

Phenols are harmful to organisms at low concentrations and classified as hazardous pollutants because of their potential to harm human health. They exist in different concentrations in wastewaters originated from coking, synthetic rubber, plastics, paper, oil, gasoline, etc. Biological treatment, activated carbon adsorption and solvent extraction are some of the most widely used methods for removing phenol and family compounds from wastewaters [73-76]. Biological treatment is economical, practical, promising and versatile approach for it leads to complete mineralization of phenol. Many aerobic bacteria are capable of using phenol as the sole source of carbon and energy [77]. In recent years, the strain of *Pseudomonas putida* has been the most widely used to degrade phenol. Under aerobic conditions, phenol may be converted by the bacterial biomass to CO_2; other intermediates such as benzoate, catechol, *cis*-cis-muconate, β-ketoadipate, succinate and acetate are formed during the biodegradation process [77, 78]. *p*-Nitrophenol (PNP) is one of the most widely used nitrophenolic compounds in industry and finds important applications in agriculture, polymers, pigment and pharmaceutical industries. However, PNP is highly toxic for both the environment and humans and its efficient removal from the environment is required. Hydroquinone (HQ), 4-nitrocatechol (4-NC) and 1,2,4-benzenetriol (1,2,4-BT) are the metabolic intermediates of the PNP biodegradation [80,81].

Chlorinated phenols are common and encountered even in relatively pristine environments [82,83]. These compounds are formed during the bleaching of pulp with chlorine [82-84]. As the pulp accounts only for about 40-45% of the original weight of the wood, these effluents are heavily loaded with organics [85]. Chlorophenols are also used as fungicides and may be formed from hydrolysis of chlorinated phenoxyacetic acid herbicides. Chlorophenols, part of the *adsorbable organic halides* (AOX), are present in bleaching effluents at concentrations ranging from 0.1 to 2.6 ppm [86]. Aqueous effluents from industrial operations such as polymeric resin production, oil refining and coking plants also contain chlorophenolic compounds. Pentachlorophenol (PCP) is the second most heavily used pesticide in the US. As compared to phenol, chlorophenolic compounds are more persistent in the environment. Toxicity and bioaccumulative potential of chlorophenols increases with the degree of chlorination and with chlorophenol lipophilicity. Haloaromatic compounds are degraded via the formation of halocatechols as intermediates which are subsequently cleaved by dioxygenases, by the mechanism delineated earlier. Dehalogenation then occurs by the elimination of the hydrogen halide, with subsequent double bond formation on the aliphatic intermediate [87]. In anaerobic environments, the biodegradation of chlorinated aromatics takes place through reductive dehalogenation leading to the formation of less toxic and more biodegradable compounds. Reductive dechlorination of 2,4-dichlorophenol is followed by carboxylation, ring fission and acetogenesis, and methanogenesis which finally led to the complete mineralization of 2,4-DCP, which is also biodegraded to 4-chlorophenol in anaerobic sediments. Similarly, biodegradation of PCP under anaerobic conditions occurs through reductive dechlorination [88].

5.4. Fluoro benzenes

Toluene degrading enzymes can transform many 3-fluoro-substituted benzenes to the corresponding 2,3-catechols with the concomitant release of inorganic fluoride. The substrates that induce 2,3-dioxygenase are 3-fluorotoluene, 3-fluorotrifluorotoluene, 3-flurohalobenzene,

3-fluoronisole, and 3-fluorobenzonitrile. While 3-fluorotoluene and 3-fluoronisole produce only deflorinated catechols, other substrates led to catechol products both with and without the toluene substituent [89].

5.5. Polycyclic Aromatic Hydrocarbons (PCAHs)

PCAHs are toxic, mutagenic and resist biodegradation [90]. Many strategies have been developed to treat them, including volatilization, photooxidation, chemical oxidation, bioaccumulation, and adsorption on soil particles [91]. Soil clean-up may be achieved using different remediation technologies, among which bioremediation is an effective and low-cost alternative that has garnered widespread use [92]. Two processes have been found to increase the activity of microorganisms during bioremediation: bio-stimulation and bio-augmentation. Bio-stimulation involves the addition of nutrients and/or a terminal electron acceptor to increase the meager activity of indigenous microbial populations. Bio-augmentation involves the addition of external microbial strains (indigenous or exogenous) which have the ability to degrade the desired toxic compounds [93]. The added specific PCAHs degrading strain, which has a competitive capacity to become dominant species with indigenous microbial strains or grow simultaneously with indigenous microbial strains, may greatly enhance the rate of PCAHs biodegradation [94,95]. The ability to degrade PCAHs depends on the complexity of their structure and the extent of enzymatic adaptation by bacteria. In general PCAHs with 2 or 3 aromatic rings are readily degraded, but those with 4 or more are usually recalcitrant and genotoxic. Such examples of PCAHs are acenaphthene, acenaphthylene, anthracene, naphthalene, fluorene, phenanthrene, chrysene, pyrene, etc. The major metabolites are 4-phenanthroic acid and 4-hydroxyperinapthenone. Cinnamic and phthalic acids are ring fission products [96].

Naphthalene is carcinogenic and persistent organic pollutant [97]. Bacteria such as *Pseudomonas putida, Rhodococcus opacus, Mycobacterium* sp., *Nocardia otitidiscaviarum*, and *Bacillus pumilus* are known to biodegrade naphthalene [98-102]. Some metabolites of naphthalene, such as salicylic acid, 1-naphthol and *o*-phthalic acid could be degraded and support cell growth (Scheme 4). Phenanthrene was used as a model compound for PCAH degradation which shows 1-hydroxy 2-naphthoic acid (1H2NA) as intermediate biodegradation product [103].

5.6. Plasticizers

Plasticizers are polymeric additives, used to impart flexibility to polymer materials. The biodegradation of some plasticizers can lead to the formation of metabolites with increased persistence and toxicity relative to the original compounds [104-106]. Use of plasticizers has grown considerably, both with respect to product variety and production volume [107]. Phthalates are the most widely used plasticizers. Presence of phthalates and their metabolites in rats, mice, human plasma and liver are related to adverse health effects such as endocrine disruption and peroxisome proliferation [108,109]. The high production volumes of phthalates and their incomplete biodegradation have led to the presence of these compounds and a number of toxic and stable metabolites in surface waters, groundwater, air, soil and tissue of living organisms [104, 110-113]. Such findings have led to stricter environmental regulations

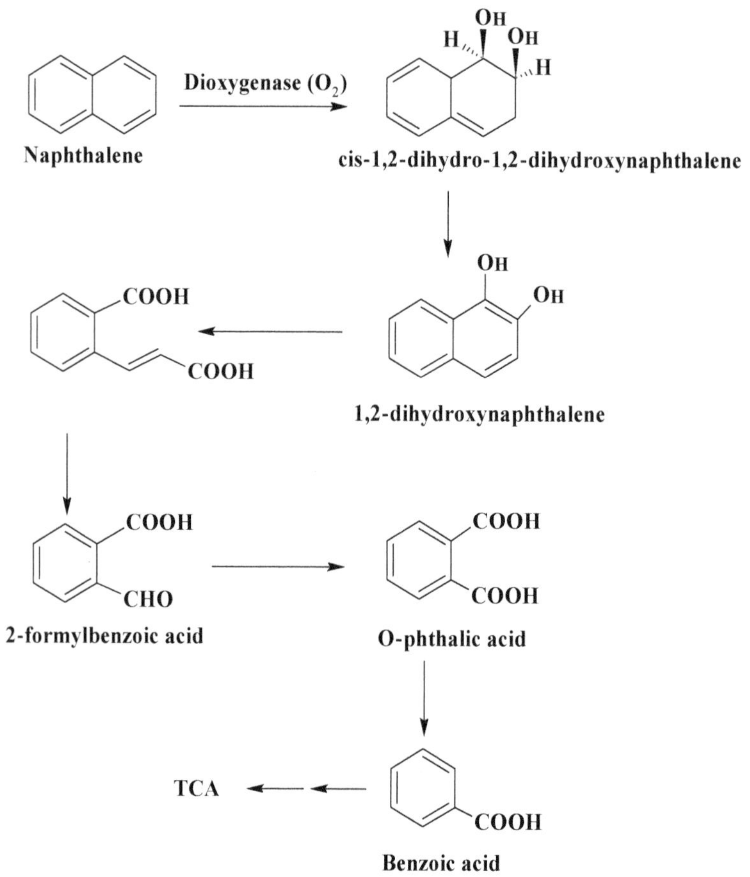

Scheme 4. Proposed pathway for the degradation of naphthalene [103]

and consequently have broadened the criteria used to evaluate plasticizers. Consequently, dibenzoates have been approved by the European Chemical Agency as alternatives to phthalates [114]. However, the degradation of dipropylene glycol dibenzoate (D(PG)DB) and diethylene glycol dibenzoate (D(EG)DB) by common soil microorganisms such as *Rhodotoru-la rubra* and *Rhodococcus rhodochrous* can lead to the formation and accumulation of monoben-zoate metabolites [115,116] that exhibit high acute toxicity [115]. Other compounds including 1,5-pentandiol and 1,6- hexanediol dibenzoates were reported to produce less stable metabo-lites and have also been tested as potential alternatives to commercial dibenzoate plasticizers [116-118]. Scheme 5 shows the biodegradation products of dibenzoates by *R. Rhodochrous*, which include 2-[2-(benzoyloxy)propoxy] propanoic acid, 1,3-propanediol monobenzoate and 3-(benzoyloxy) propanoic acid [119].

5.7. Plastics

Over the years, plastics have brought economic, environmental and social advantages. Today's material world uses tremendous quantities of plastics of all hue and origins. However, their wide spread use has also increased plastic waste, which brings its own economic, environmental and social problems. The redesign of plastic products, whether individual polymer or product structure, could help alleviate some of the problems associated with plastic waste. Redesign could have an impact at all levels of the hierarchy established by the European Waste Framework Directive: prevention, re-use, recycle, recovery and disposal [120].

Polyethylene, polypropylene and polystyrene, and water-soluble polymers, such as polyacrylamide, polyvinyl alcohol and polyacrylic acid are bulk polymers used in a variety of industries and products. Some of the plastics are not biodegradable and deleterious to the environment due to their accumulation. Plastics can be disposed of by incineration or recycling, but incineration is very difficult, dangerous and expensive whereas recycling is a long process and not very efficient. Some plastics still cannot be recycled or incinerated due to pigments, coatings and other additives during manufacture of materials. Making biodegradable and ecofriendly plastics will avoid accumulation, recycling and incineration [121].

5.7.1. Polyvinyl alcohol

Polyvinyl alcohol (PVA) is water-soluble but also has thermoplasticity. In addition to its use as a water-soluble polymer, for instance, as a substituent for starch in industrial processes, it can also be molded in various shapes, such as containers and films. PVA can therefore be used to make water-soluble and biodegradable carriers, which may be useful in the manufacture of delivery systems for chemicals such as fertilizers, pesticides, and herbicides. Among the vinyl polymers produced industrially, PVA is the only one known to be mineralized by microorganisms [122]. Extensive use of PVA, in textile and paper industries generates considerable amount of contaminated wastewaters [121]. In aqueous solution, the biodegradation mechanism of PVA involves the random endocleavage of the polymer chains. The initial step is associated with the specific oxidation of methane-carbon bearing the hydroxyl group, as mediated by oxidase and dehydrogenase type enzymes, to give β-hydroxyketone as well as 1,3-diketone moieties. The latter groups are able to facilitate the carbon-carbon bond cleavage as promoted by specific β-diketone hydrolase, leading to the formation of carboxylic and methyl ketone end groups [123,124]. Most of the PVA-degrading microorganisms are aerobic bacteria belonging to *Pseudomonas*, *Alcaligenes*, and *Bacillus* genus. A very moderate PVA biodegradation was reported [125-128].

5.7.2. Polyhydroxyalkanoates

Polyhydroxyalkanoates (PHAs) are degraded to CO_2 and water in aerobic conditions and methane in anaerobic conditions by microbes found in soil, water and other various natural habitats. PHAs are the only proposed replacement polymers that are completely biodegradable [129]. The structures of these polymers have a very similar structure of the petroleum-derived thermoplastics [130].

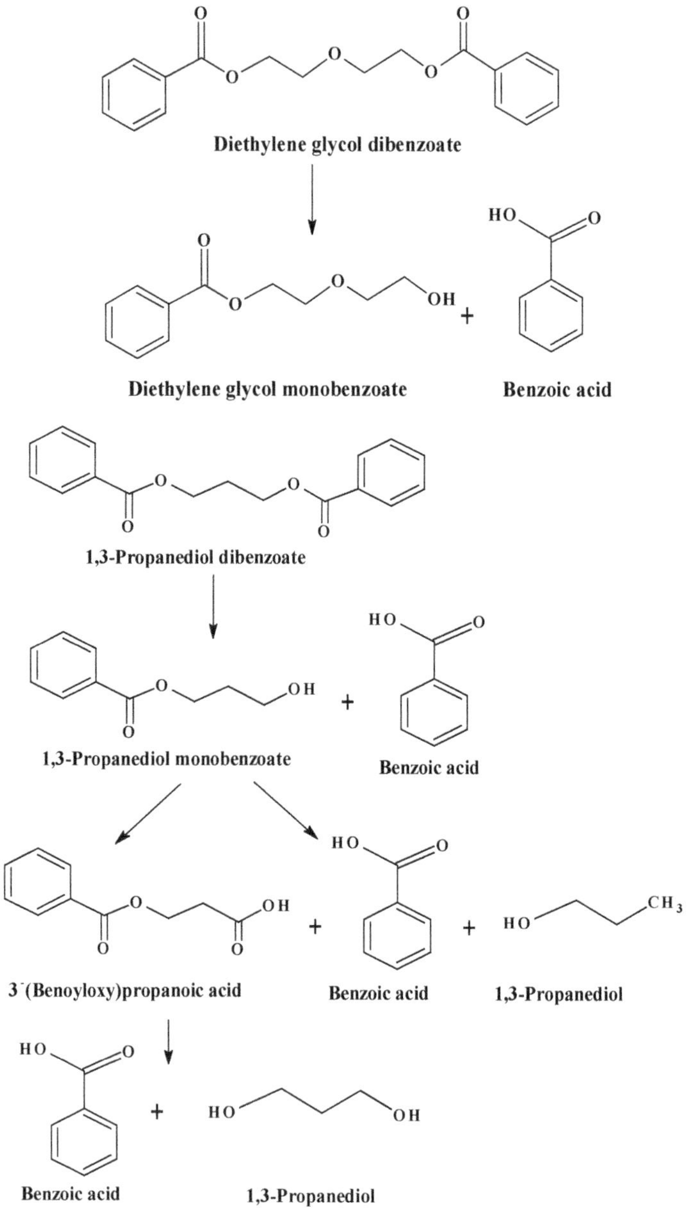

Scheme 5. Proposed biodegradation pathways of diethylene glycol dibenzoate and 1,3-propanediol dibenzoate [116]

PHAs also possess similar physical properties as plastics including the ability to be molded, made into films, and also into fibers. Efforts are underway to identify bacteria, which produce various kinds of PHAs [129] as well as the production of these polyesters or create certain kinds of PHAs by changing the kind of bacteria [130] and/or the substrates given to the bacteria and genetically enhancing bacteria [131].

6. Prospective of anaerobic digestion and biogas energy

The foregoing analysis shows that anaerobic digestion technologies have matured so far to treat several organic micro-pollutants, halogenated compounds, substituted aromatics, azo-linkages, nitro-aromatics and the like in industrial effluents and also for municipal effluents containing industrial loads. A very high growth potential is envisaged for the anaerobic digestion of organic fraction of municipal solid waste [27]. Novel reactor and control systems ought to be developed for different purposes depending on the source of pollutants or biomass. Anaerobic digestion of sewage sludge followed by recycling on agricultural land is currently the largest world-wide application of anaerobic processes. Treatment of sludge and slurries targeted at the production of safe end products can be tackled with niche anaerobic technologies [16]. It is predicted that major future process developments will come from the deployment of pre- and post treatment processes, including physical, chemical and biological processes, for the reclamation of the products from the wastewater treatment system. Wastewater treatment for reuse will be effective if anaerobic digestion is adopted for mineralizing organic matter. Hence, anaerobic digestion has the potential to play a major role in closing water, raw materials, and nutrient cycles in industrial processes [37]. Further development is required on the community on-site treatment of domestic sewage under a wide range of conditions, opting for the reuse of the treated water in agriculture and making use of the mineralized nutrients for fertilization purposes. An upstream integration of the anaerobic process with industrial primary production processes under extreme conditions of temperature, pH, salinity, toxic and recalcitrant compounds, and variable load is envisaged in future [39].

There is an emphasis worldwide on renewable energy system among which biogas produced from any biological feedstocks including primary agricultural sectors and from various organic waste streams will come in to prominence in near future [22]. It is estimated [3] that at least 25% of all bioenergy in the future can originate from biogas, produced from wet organic materials like animal manure, slurries from cattle and pig production units as well as from poultry, fish and fur, whole crop silages, wet food and feed wastes, etc. Anaerobic digestion of animal manure offers several environmental, agricultural and socio-economic benefits throughout such as improved fertilizer quality of manure, considerable reduction of odors and inactivation of pathogens and more importantly production of biogas production, as clean, renewable fuel, for multiple utilizations [16]. This biogas can be upgraded to natural gas to mix with the existing natural gas grid which will be cost effective. The potential development of biogas from manure co-digestion includes the use of new feedstock types such as by-products from food processing industries, bio-slurries from biofuels processing industries as well as the biological degradation of toxic organic wastes from pharmaceutical industries, etc. [3,16,22]. This will also call for better reactor systems and careful process control to increase the biogas yield, which will be more attractive if coupled with less capital and operating costs. Integration of biogas production into the national energy grids will eventually be commercially viable since the biogas from anaerobic co-digestion of animal manure and suitable organic wastes would overcome the major environmental and veterinary problems of the animal production and organic waste disposal.

7. Plastic waste separation, reprocessing and recycle

In 2009, around 230 million tonnes of plastic was produced; ~25 % which was used in the European Union [131]. About 50 % plastic is used for single-use disposable applications, such as packaging, agricultural films and disposable consumer items [132]. Although plastics consume approximately 8 % world oil production: 4 % as raw material for plastics and 3-4 % as energy for manufacture [132], supplies are being depleted. Bioplastics make up only 0.1 to 0.2 % total plastics [115]. It is estimated that plastics reduce 600 to 1300 million tonnes of CO_2 through the replacement of less efficient materials, lighter and fuel efficient vehicles, housing sector, contribution to insulation, preservation of food, packaging, use in wind power rotors and solar panels [133]. However, plastic littering and pollution of land and sea have been matters of great concern which should be strategically and technologically solved. Plastics recovery, in addition to increased diversion from disposal, results in significant energy savings (~50-75 MBtu/ton of material recycled) compared with the production of virgin materials, which also leads to reductions in greenhouse gas emissions due to avoided fuel use. Limiting the plastics that enter landfills can lower the costs associated with waste disposal. It is believed that the recycled plastic will fetch as much as 70 % of the typical price for virgin plastics [136].

7.1. Waste reduction hierarchy

The motto of waste reduction by plastics is by following the principles of (i) prevention, (ii) reuse (iii) recycle, (iv) recovery, and (v) disposal [119].

i.   *Prevention* – Using minimum and as less types of plastic in the product by clever product redesign.

ii.   *Reuse* – Products could be designed for reuse by facilitating the dismantling of products and replacement of parts. This could involve standardizing parts across manufacturers [137].

iii.   *Recycle* – Some types of plastics are easier to recycle than others, for example poly-ethylene terephthalate (PET). By using fewer types and colors (or colorless) of plastics the recycling process becomes easier. The use of "intelligent" or smart polymers that undergo changes under certain conditions could also reduce disassembly time [138]. For example, a polymer that changes shape when subject to magnetic or electric fields could aid the disassembly of electronic goods.

iv.   Recovery – Energy can be recovered from plastics in waste-to-energy plants. By designing products to consider the possibility of energy recovery, plastic may have a greater end-of-life use.

v.   *Disposal* – Biodegradable plastics are less persistent in the environment than traditional plastics, but need specific and suitable end-of-life treatment.

7.2. Bioplastics

Since disposal is one of the important aspect, bioplastics are being favored. There are three main categories of bio-based plastics: (i) Natural polymers from renewable sources, such as cellulose, starch and plant-based proteins, (ii) Polymers synthesised from monomers derived from renewable resources. For example, polylactic acid (PLA) is produced by the fermentation of starch, corn or sugar, (iii) Polymers produced by microorganisms. For example, PHA (polyhydroxyalkanoate) is produced by bacteria through fermentation of sugar or lipids [139].

Biodegradable plastics are not by definition bio-based and bio-based plastics are not always biodegradable, although some fall into both categories, such as PHAs. The term *bioplastics* is often used to refer to both bio-based and biodegradable plastics. The main applications of bioplastics are disposable plastic bags, packaging and loose fill packaging (beads and chips), dishes and cutlery, electronic casings and car components. However, bioplastics cannot substitute all types of plastic; particularly certain types of food packaging that require gas permeability [135]. Development of novel biodegradable plastic is a solution for the plastic disposal problem since plastics are immiscible in water and are thermo-elastic polymeric materials. Biodegradability of plastics is governed by both their chemical and physical properties. Other factors affecting degradability are the forces associated with covalent bonds of polymer molecules, hydrogen bonds, van der Waals forces, coulombic forces, etc. Enzymatic degradation is an effective way. Lipase and esterase can hydrolyze fatty acid esters, triglycerides and aliphatic polyesters. These lipolytic enzymes have an important role in the degradation of natural aliphatic polyesters such as cutin, suberin and esteroid in the natural environment and animal digestive tract.

As stated earlier, biodegradable plastics decompose in the natural environment from the action of bacteria. Biodegradation of plastics can be achieved through the action of micro-bacteria and fungi in the environment to metabolize the molecular structure of plastic films to produce an inert humus-like material that is less harmful to the environment, along with water, carbon dioxide and/or methane. They may be composed of either bioplastics or petro-plastics. The use of bio-active compounds compounded with swelling agents ensures that, when combined with heat and moisture, they expand the plastic's molecular structure and allow the bio-active compounds to metabolize and neutralize the plastic [140]. Compostable plastics are biodegradable and meet certain criteria, such as rate of biodegradation and impact on the environment. Degradable plastics include biodegradable and compostable plastics, but also plastics that degrade by chemical and physical processes such as the action of sunlight. Purely biodegradable plastics are different from oxy-biodegradable plastics, which contain small amounts of metal salts to speed up degradation. It has been suggested that this process be called "oxo-fragmentation" to avoid confusion [139,140].

It is possible to produce polymers biologically, e.g., PHA grown in genetically modified corn plant leaves, PLA (polylactic acid) produced by the fermentation of sugars extracted from plants, PHA produced by bacteria. Bioplastics could also help alleviate climate change by reducing the use of petroleum for the manufacture of traditional plastics. It is claimed that CO_2 emissions released at the end-of-life of bio-based plastics are offset by absorption of CO_2 during the growth of plants for their production [141].

7.3. Sorting plastic materials

The technical difficulties and high cost associated with separating plastics have limited recycling in the past. Consumer goods often contain as many as 20 different types of plastic as well as non-plastic materials such as wood, rubber, glass, and fibers. There is upsurge of new products and pigment types, which can pose a challenge to the recycling infrastructure. Consequently, the cost of producing virgin materials is often less than the cost of collecting and processing post-consumer plastics. Used plastic material will contain more than one base polymer, and resins with a variety of additives, including coloring agents and thus technologies for cleaning and separating the materials are an important part of most plastics recycling systems. A particular concern for recycled plastics is their use as food containers requiring stringent regulations to avoid contamination [140].

Separation of different types of polymers from each other is many times a desired part of plastics recycling processes which are classified as macrosorting, microsorting, or molecular sorting.

7.3.1. Macrosorting

Macrosorting involves the sorting of whole or nearly whole objects such as separation of PVC bottles or caps from PET bottles, separation of polyester carpet from nylon carpet, and sorting of automobile components by resin type. Various devices are now commercially available to separate plastics by resin type, which typically rely on differences in the absorption or transmission of certain wavelengths of electromagnetic radiation, or color or resin type. Particularly for recycling of appliances, carpet, and automobile plastics, several IR spectra based equipment are used [135].

7.3.2. Microsorting

Microsorting is a size-reduction process to reduce the plastic material in to small pieces which is then separated by resin type or color; for instance, separation of high-density polyethylene (HDPE) base cups from PET soft drink bottles using a sink-float tank. More modern separation processes, such as the use of hydrocyclones, also rely primarily on differences in the density of the materials for the separation. A number of other characteristics have also been used as the basis for microsorting systems, including differences in melting point and in triboelectric behavior. In many of these systems, proper control over the size of the plastic flakes is important in being able to reliably separate the resins. Some systems rely on differences in the grinding behavior of the plastics combined with sieving or other size-based separation mechanisms for sorting. Sometimes cryogenic grinding is used to facilitate grinding and to generate size differences [135].

Three new separation technologies, developed by MBA Polymers, Argonne National Laboratory, and Recovery Plastics International (RPI), could break down these barriers and increase plastics recycling [138].

7.3.2.1. Automated separation

According to the process developed by MBA Polymers, plastic scraps from computers and other electronics are first ground into small pieces. Magnets and eddy-current separators then extract ferrous and non-ferrous metals. Paper and other lighter materials are removed with jets of air. Finally, a proprietary sorting, cleaning, and testing process involving various technologies, allows the separation of different types of plastics and compound them into pelletized products comparable to virgin plastics [138].

7.3.2.2. Froth flotation

Argonne National Laboratory (ANL) has developed a process to separate acrylonitrile-butadiene styrene (ABS) and high-impact polystyrene (HIPS) from recovered automobiles and appliances. The froth flotation process separates two or more equivalent-density plastics by modifying the effective density of the plastics. There is a careful control of the chemistry of the aqueous "froth" so that small gas bubbles adhere to the solid surface and facilitate the plastic to float to the top [135].

7.3.2.3. Skin flotation

Recovery Plastics International (RPI) has developed an automated process capable of recovering up to 80 % plastics found in automobile shredder residue (ASR). The process starts with the separation of light lint materials, followed by the removal of rocks and metals, granulation, washing, and, finally, automated skin flotation separation. This final step adds a skin of plasticizer to make the surface of the targeted plastic hydrophobic. Air bubbles then can attach to the plastic, allowing it to float above denser, uncoated pieces. It is estimated this new skin flotation technology could be used to treat about one-third of the estimated 7 million tons of ASR disposed off each year [141].

7.3.3. Molecular sorting

Molecular sorting deals with sorting of materials whose physical form has been completely disrupted, such as by dissolving the plastics in solvents using either a single solvent at several temperatures or mixed solvents, followed by reprecipitation. There is a need to control emissions and to recover the solvents, without any residual solvent in the recovered polymer to avoid leaching in stored material. There are at present no commercial systems using this approach. Some research effort has focused on facilitating plastics separation by incorporating chemical tracers into plastics, particularly packaging materials, so that they can be more easily identified and separated.

It has become obvious that many of the difficulties of recycling plastics are related to difficulties in separating plastics from other wastes and in sorting plastics by resin type. Design of products can do a lot to either aggravate or minimize these difficulties [134,135]. The concept of green product embeds recycling at the design stage itself.

7.4. Plastic reprocessing and recycling

For plastics recycling to be effective, it is necessary to have (i) a system for collecting the targeted materials, (ii) a facility capable of processing the collected recyclables into a form which can be utilized to make a new product and, (iii) new products made in whole or part from the recycled material must be manufactured and sold.

Processing of recyclable plastics is necessary to transform the collected materials into raw materials for the manufacture of new products. Three general categories of processing can be identified: (1) physical recycling, (2) chemical recycling, and (3) thermal recycling, wherein the particulars of the processing are often specific to a given plastic or product.

7.4.1. Physical processes

Physical recycling, the most popular option, covers size and shape alteration, removing contaminants, blending in additives if desired, and similar approaches that change the appearance of the recycled material, but do not alter its basic chemical structure. Plastic containers, for example, are processed including grinding, air classification to remove light contaminants, washing, gravity-separation, screening, rinsing, drying, and often melting and pelletization, accompanied by addition of colorants, heat stabilizers, or other ingredients, depending on type of plastic [132].

7.4.2. Chemical reactions

Chemical recycling of plastics deals with chemical reactions using catalysis or solvents such as methanol, glycols or water leading to depolymerization or breaking polymers into monomers or useful chemicals, or fuels [134]. The products of the reaction then can be separated and reused to produce either the same or a related polymer. An example is the glycolysis process sometimes used to recycle polyethylene terephthalate (PET), in which the PET is broken down into monomers, crystallized, and repolymerized. Condensation polymers, such as PET, nylon, and polyurethane, typically much more amenable to chemical recycling than are addition polymers such as polyolefins, polystyrene, and polyvinyl chloride (PVC). Most commercial processes for depolymerization and repolymerization are restricted to a single polymer, which is usually PET, nylon 6, or polyurethane. Methanolysis is another common reaction using methanol [134].

7.4.3. Thermal cracking

Thermal cracking or recycling also involves cracking of the chemical structure of the polymer using heat in the absence of sufficient oxygen for combustion. At these elevated temperatures, the polymeric structure breaks down. Thermal recycling can be applied to all types of polymers. However, the typical yield is a complex mixture of products, even when the feedstock is a single polymer resin. If reasonably pure compounds can be recovered, products of thermal recycling can be used as feedstock for new materials. When the products are a complex mixture which is difficult to separate, they are most often used as fuel. There are relatively few commercial operations today which involve thermal recycling of plastics. Some European nations have such feedstock recycling facilities. Many plastic resin companies use fluidized bed cracking to produce a waxlike material from mixed plastic waste [134-136, 139]. The product, when mixed with naptha, can be used as a raw material in a cracker or refinery to produce feedstocks such as ethylene and propylene. In certain case, syn gas can be produced and used in Fisher-Tropsch synthesis to produce valuable chemicals.

In landfill, both synthetic and naturally occurring polymers do not get the necessary exposure to UV and microbes to degrade. The discarded plastics occupy space and none of the energy put into making them is being reclaimed. Reclaiming the energy stored in the polymers can be done through incineration, but this can cause environmental damage by release of toxic gases into the atmosphere. Therefore, recycling is a viable alternative in getting back some of this energy in the case of some polymers. With ever increasing petroleum prices, it would be financially viable to recycle polymers rather than produce them from raw materials [141].

8. Conclusions

The modern society needs thousands of chemicals and materials of all sorts which are produced annually and used in all sectors of economy. However, their fate in the environment is of great concern since some are toxic, recalcitrant and bioacumulating and hence their discharge into the environment must be regulated. Better understanding of the mechanism of biodegradation has a high ecological significance that depends on indigenous microorganisms to transform or mineralize the organic contaminants. Thus, biodegradation processes differ greatly depending on conditions, but frequently the main final products of the degradation are carbon dioxide and/or methane. Microorganisms have enzyme systems to degrade and utilize different hydrocarbons as a source of carbon and energy. Slow and partial biodegradation of chlorophenolic compounds under aerobic as well as anaerobic natural environment has been observed. Aerobic degradation takes place via formation of catechols while anaerobic degradation occurs via reductive dechlorination. Acclimatization of biomass to chlorophenols markedly enhances their ability to degrade such compounds, both by reducing the initial lag phase as well as by countering biomass inhibition. Aerobic processes as well as anaerobic processes partially remove chlorophenols. However, enhanced removal efficiency can be obtained by operating anaerobic and aerobic treatment processes in combination. Thus microbial degradation can be a key component for clean-up strategy of organopollutants and plastics.

Renewable energy system among which biogas produced from biological feedstocks will play a major role in energy sector. Anaerobic digestion of animal manure, slurries from cattle and pig production units as well as from poultry, fish and fur, whole crop silages, wet food and feed wastes, etc offers several environmental, agricultural and socio-economic benefits by improved fertilizer quality of manure, considerable reduction of odors, inactivation of pathogens and production of biogas production, as clean and renewable fuel. This biogas can be upgraded to natural gas to inject in to the existing natural gas grid which will be cost effective. Biogas from anaerobic co-digestion of animal manure and suitable organic wastes would overcome the major environmental and veterinary problems of the animal production and organic waste disposal.

The recycling of plastics is environmentally beneficial because plastics reduce millions of tonnes of CO_2 emissions through the replacement of less efficient materials, development of lighter and fuel efficient transport systems, housing material, energy saving insulation, food preservation and storage, energy efficient packaging, use in wind power rotors and solar panels. Processing of recyclable plastics is necessary to transform the collected materials into raw materials for the manufacture of new products. Bioplastics offer a very good solution to environmentally deleterious materials. Biodegradation of plastics can be achieved through the action of micro-bacteria and fungi.

Author details

Ganapati D. Yadav* and Jyoti B. Sontakke

*Address all correspondence to: gdyadav@yahoo.com; gd.yadav@ictmumbai.edu.in

Department of Chemical Engineering, Institute of Chemical Technology, Matunga, Mumbai, India

References

[1] Alexander M. Biodegradation and Bioremediation. Academic Press: New York; 1999.

[2] Lily Y, Young LY, Cerniglia CE. Microbial Transformation and Degradation of Toxic Organic Chemicals. Wiley-Liss Inc. New York, NY; 1995.

[3] Holm-Nielsen JB, Al Seadi T, Oleskowicz-Popiel P. The future of anaerobic digestion and biogas utilization. Bioresource Technology 2009; 100: 5478–5484.

[4] Steinfeld H, Gerber P, Wasenaar T, Castel V, Rosales M, de Haan C. Livestock's long shadow. Environmental Issues and Options. Food and Agriculture Organization (FAO) of United Nations; 2006.

[5] Nielsen LH, Hjort-Gregersen K, Thygesen P, Christensen J. Samfundsøkonomiske analyser af biogasfllesanlg. Rapport 136; 2002. Fødevareøkonomisk Institut, København (Summary in English).

[6] http://www.unesco.org/new/en/natural-sciences/ (accessed December 2012)

[7] Van Agteren MH, Keuning S, Janssen D, Handbook on Biodegradation and Biological Treatment of Hazardous Organic Compounds. Kluwer, Dordrecht, The Netherlands; 1998.

[8] Tokiwa Y, Calabia BP, Ugwu CU, Aiba S. Biodegradability of Plastics. International Journal of Molecular Science 2009; 10: 3722–3742.

[9] Griffin GJL. Chemistry and Technology of Biodegradable Polymers, Springer, London; 1994.

[10] Schmidt M., editor. Synthetic Biology: Industrial and Environmental Applications, Wiley-Blackwell; 2012.

[11] NIIR Board of Consultants and Engineers (Ed.), Medical, Municipal and Plastic Waste Management Handbook. National Institute of Industrial Research, New Delhi; 2009.

[12] Stuart PR, El-Halwagi MM., editor. Integrated Biorefineries: Design, Analysis and Optimization, CRC Press; 2012.

[13] Perez JJ, Munoz-Dorado J, de la Rubia TJ, Martınez J. Biodegradation and biological treatments of cellulose, hemicelluloses and lignin: an overview. International Microbiology 2002; 5: 53-63.

[14] Berna JL, Cassani G, Hager CD, Rehman N, Lopez I, Schowanek D, Steber J, Taeger K, Wind T. Anaerobic Biodegradation of Surfactants-Scientific Review. Tenside Surfactants Detergents 2007; 44: 312-347.

[15] Fritsche W, Hofrichter M. Aerobic Degradation by Microorganisms: Principles of Bacterial Degradation. In: Rehm HJ, Reed G, Puhler A, Stadler A. (eds.) Biotechnology, environmental processes II, vol IIb. Wiley-VCH, Weinhein. p145-167.

[16] Lier JB van, Tilche A, Ahring BK, Macarie H, Moletta R, Dohanyos M, Hulshoff Pol LW, Lens P, Verstraete W. New perspectives in anaerobic digestion. Water Science and Technology 2001; 43(1): 1-18.

[17] Rozzi A, Remigi E. Anaerobic biodegradability: Conference Proceeding. In: 9th World Congress, Anaerobic digestion 2001, Workshop 3 Harmonisation of anaerobic activity and biodegradation assays, 9-2-2001, Belgium.

[18] Dolfing J, Bloemen GBM. Activity measurement as a tool to characterize the microbial composition of methanogenic environments. Journal of Microbiological Methods 1985; 4: 1-12.

[19] Soto M, Mendez R, Lema JM. Methanogenic activity tests. Theoretical basis and ex-
      perimental setup. Water Research 1993; 27: 850–857.

[20] Angelidaki I, Ahring BK. Thermophilic anaerobic digestion of livestock waste: the ef-
      fect of ammonia. Applied Microbiology Biotechnology 1993; 38: 560–564.

[21] Angelidaki I. Sanders W. Assessment of the anaerobic biodegradability of macropol-
      lutants. Reviews in Environmental Science and Biotechnology 2004; 3: 117–129.

[22] Holm-Nielsen JB, Oleskowicz-Popiel P, 2007. The future of biogas in Europe: Visions
      and targets until 2020. In: Proceedings: European Biogas Workshop-Intelligent Ener-
      gy Europe, 14–16 June 2007, Esbjerg, Denmark. Mata-Alvarez J, Macé S, Llabrés P,
      Anaerobic digestion of organic solid wastes. An overview of research achievements
      and perspectives. Bioresource Technology 2000; 74: 3-16.

[23] Reusser S, Zelinka G. Laboratory-scale comparison of anaerobic-digestion alterna-
      tives. Water Environment Research 2004; 76(4): 360-379.

[24] David C, Inman DC. Comparative studies of alternative anaerobic digestion technol-
      ogies. M.S. (Environ. Eng.) Thesis, Virginia Polytechnic Institute and State Universi-
      ty; 2004.

[25] Riggle D. Acceptance improves for large-scale anaerobic digestion. Biocycle 1998; 39
      (6): 51-55.

[26] Mata-Alvarez, J., Tilche, A., Cecchi, F., editor. The treatment of grey and mixed solid
      waste by means of anaerobic digestion: future developments. Proceedings of the Sec-
      ond International Symposium on Anaerobic Digestion of Solid Wastes, Barcelona,
      vol. 2. Graphiques 92, 15-18 June, 1999. p302-305.

[27] Christiansen N. Hendriksen HV, Jarviene KT, Ahring B. Degradation of chlorinated
      aromatic compounds in UASB reactors. Water Science and Technology 1999; 31:
      249-259.

[28] Man AWA de, Last ARM van der, Lettinga G. The use of EGSB and UASB anaerobic
      systems or low strength soluble and complex wastewaters at temperatures ranging
      from 8 to 30°C. Proceedings of the 5th International Symposium on Anaerobic Diges-
      tion. Bologna, Italy, 1988. p197-211.

[29] Driessen WJBM, Habets LHA, Groeneveld N. New developments in the design of
      Upflow Anaerobic Sludge Bed reactors. 2nd Specialised IAWQ conference on Pre-
      treatment of Industrial Wastewaters, October 16-18,1996.

[30] Zoutberg GR, Been P de. The biobed EGSB (Expanded Granular Sludge Bed) systems
      covers short comings of the upflow anaerobic sludge blanket reactors in the chemical
      industry. Water Science and Technology 1997; 35(10): 183-188.

[31]   Collins AG, Theis TL, Kilambi S, He L, Pavlostathis SG. Anaerobic treatment of low strength domestic wastewater using an anaerobic expanded bed reactor. Journal of Environmental Engineering 1998: 652-659.

[32]   Nagano A, Arikawa E, Kobayashi H. Treatment of liquor wastewater containing high strength suspended solids by membrane bioreactor system. Water Science and Technology 1992; 26(3-4): 887-895.

[33]   Garuti G, Dohanyos M, Tilche A. Anaerobic-aerobic wastewater treatment system suitable for variable population in coastal are: the ANANOX process. Water Science and Technology 1992; 25(12):185-195.

[34]   Oude Elferink SJWH, Visser A, Hulshoff Pol LW, Stams AJM. Sulfate reduction in methanogenic bioreactors. FEMSMicrobiology Reviews 1994; 15: 119-136.

[35]   Boopathy R, Kulpa CF, Manning J. Anaerobic biodegradation of explosives and related compounds by sulfate-reducing and methanogenic bacteria: a review. Bioresource Technoology 1998; 63(1): 81-89.

[36]   Houten RT van, Lettinga G. Biological sulphate reduction with synthesis gas: microbiology and technology. In: Wijffels RH., Buitelaar RM., Bucke C., Tramper J. (eds.) Progress in Biotechnology. Elsevier, Amsterdam, The Netherlands; 1996. pp. 793-799..

[37]   Jetten MSM, Strous M, Pas-Schoonen KT Van de, Schalk J, Van Dongen UGJM, Van De Graaf AA, Logemann S, Muyzer G, Van Loosdrecht MCM, Kuenen JG. The anaerobic oxidation of ammonium. FEMS Microbiology Reviews 1999; 22: 421-437.

[38]   Lier JB van, Lettinga, G. Appropriate technologies for effective management of industrial and domestic wastewaters: the decentralised approach. Water Science and Technology 1999; 40 (7): 171-183.

[39]   Abrahamsson K, Klick S. Degradation of Halogenated Phenols in Anoxic Marine Sediments. Marine Pollution Bulletin 1991; 22: 227-233.

[40]   Suflita JM, Horowitz A, Shelton DR, Tiedje JM. Dehalogenation: A Novel Pathway for the Anaerobic Biodegradation of Haloaromatic Compounds. Science 1982; 218: 1115-1117.

[41]   Annachhatre AP, Gheewala SH. Biodegradation of Chlorinated Phenolic Compounds. Biotechnology Advances 1996; 14(1): 35-56.

[42]   Howard PH. Handbook of Environmental Degradation Rates. Lewis Publishers: Chelsea MI; 1991.

[43]   Zobell CE. Action of microorganisms on hydrocarbons. Bacteriological Reviews 1946; 10(1-2): 1-49.

[44]   Atlas RM. Microbial degradation of petroleum hydrocarbons: an environmental perspective. Microbiological Reviews 1981; 45(1): 180–209.

[45] Wilson JT, Leach LE, Henson M, Jones JN. In situ biorestoration as a ground water remediation technique. Ground Water Monitoring Review 1986; 6: 56–64.

[46] Leahy JG, Colwell RR. Microbial degradation of hydrocarbons in the environment. Microbiological Reviews 1990; 54(3): 305–315.

[47] Bedient PB, Rifai HS, Newell CJ. Ground Water Contamination: Transport and Remediation. PTR Prentice-Hall Inc. Englewood Cliffs, NJ; 1994.

[48] EPA (Environmental Protection Agency). Monitored Natural Attenuation of Petroleum Hydrocarbons. Remedial Technology Fact Sheet. EPA/600/F-98/021. Office of Research and Development, Washington, DC; May 1999.

[49] DeVaull G. Indoor vapor intrusion with oxygen-limited biodegradation for a subsurface gasoline source. Environmental Science and Technology 2007; 41(9): 3241–3248.

[50] Roggemans S, Bruce CL, Johnson PC. Vadose Zone Natural Attenuation of Hydrocarbon Vapors: An Empirical Assessment of Soil Gas Vertical Profile Data. API Technical Bulletin No. 15. American Petroleum Institute, Washington, DC, 2002.

[51] http://apiep.api.org/environment (accessed December 2012)

[52] EPA (Environmental Protection Agency). Petroleum Hydrocarbons And Chlorinated Hydrocarbons Differ In Their Potential For Vapor Intrusion. Office of Underground Storage Tanks, Washington, D.C. 20460. 2011. p1-11.

[53] www.epa.gov/oust (accessed December 2012)

[54] Belay N, Daniels L. Production of Ethane, Ethylene, and Acetylene from Halogenated Hydrocarbons by Methanogenic Bacteria. Applied Environmental Microbiology 1987; 53(7): 1604-1610.

[55] de Bruin WP, Kotterman MJJ, Posthumus MA, Schraa G, Zehnder AJB. Complete Biological Reductive Transformation of Tetrachloroethene to Ethane. Applied Environmental Microbiology 1992; 58(6): 1996-2000.

[56] Bradley PM, Chapelle FH. Methane as a Product of Chloroethene Biodegradation under Methanogenic Conditions. Environmental Science Technology 1999; 33(4): 653-656.

[57] Vogel TM, McCarty PL. Biotransformation of Tetrachloroethylene to Trichloroethylene, Dichloroethylene, Vinyl Chloride, and Carbon Dioxide under Methanogenic Conditions. Applied Environmental Microbiology 1985; 49: 1080-1083.

[58] Little, CD, Palumbo AV, Herbes SE, Lidstrom ME, Tyndall RL, Gilmer PJ. Trichloroethylene Biodegradation by a Methane-Oxidizing Bacterium. Applied Environmental Microbiology 1988; 54(4): 951-956.

[59] Tsien H, Brusseau GA, Hanson RS, Wackett LP. Biodegradation of Trichloroethylene by Methylosinus trichosporium OB3b. Applied Environmental Microbiology 1989; 55(12): 3155-3161.

[60] Wilson JT, Wilson BH. Biotransformation of Trichloroethylene in Soil. Applied Environmental Microbiology 1985; 49(1): 242-243.

[61] Pfaender FK. Biological Transformations of Volatile Organic Compounds in Groundwater. In: Ram NM, Christman RF, Cantor KP (eds.) Significance and Treatment of Volatile Organic Compounds in Water Supplies. Lewis Publishers: Chelsea, MI 1990. p205–226.

[62] Bouwer EJ. Bioremediation of Organic Contaminants in the Subsurface. In: Mitchell R. (Eds.) Environmental Microbiology. John Wiley & Sons: New York 1992. p287–318.

[63] Lorah MM, Olsen LD, Capone DG, Baker JE. Biodegradation of Trichloroethylene and Its Anaerobic Daughter Products in Freshwater Wetland Sediments. Bioremediation Journal 2001; 5(2): 101–118.

[64] Fetzner S. Bacterial degradation of pyridine, indole, quinoline, and their derivatives under different redox conditions. Applied Environmental Microbiology 1998; 49: 237–250.

[65] Shukla OP. Microbial transformation of quinoline by a Pseudomonas species. Applied Environmental Microbiology 1986; 51: 1332–1442.

[66] Kaiser JP, Feng YC, Bollag JM. Microbial metabolism of pyridine, quinoline, acridine, and their derivatives under aerobic and anaerobic conditions. Microbiological Reviews 1996; 60: 483–498.

[67] Carl B, Arnold A, Hauer B, Fetzner S. Sequence and transcriptional analysis of a gene cluster of Pseudomonas putida 86 involved in quinoline degradation. Gene 2004; 331: 177–188.

[68] Kilbane JJ, Ranganathan R, Cleveland L, Kayser KJ, Ribiero C, Linhares MM, Selective removal of nitrogen from quinoline and petroleum by Pseudomonas ayucida IGTN9m. Applied Environmental Microbiology 2000; 66: 688–693.

[69] Shukla OP. Microbiological degradation of quinoline by Pseudomonas stutzeri: the coumarin pathway of quinoline catabolism. Microbios 1989; 59: 47–63.

[70] Shukla OP. Microbiological transformation of quinoline by Pseudomonas sp. Applied Environmental Microbiology 1986; 51: 1332-1442.

[71] O'Loughlin EJ, Kehrmeyer SR, Sims GK, Isolation, characterization, and substrate utilization of a quinoline-degrading bacterium. International Biodeterioration and Biodegradation 1996; 38: 107–118.

[72] Sugaya K, Nakayama O, Hinata N, Kamekura K, Ito A, Yamagiwa K, Ohkawa A. Biodegradation of quinoline in crude oil. Journal of Chemical Technology Biotechnology 2001; 76: 603–611.

[73] Sun Q, Bai Y, Zhao C, Xiao Y, Wen D, Tang X. Aerobic biodegradation characteristics and metabolic products of quinoline by a Pseudomonas strain. Bioresource Technology 2009; 100: 5030-5036.

[74] Annadurai G, Juang R, Lee DJ. Microbial degradation of phenol using mixed liquors of Pseudomonas putida and activated sludge. Waste Manage 2002; 22: 703–710.

[75] Mohan D, Chander S. Single component and multi-component adsorption of phenols by activated carbons. Colloids and Surfaces A: Physicochemical &. Engineering Aspects 2001; 177: 183–196.

[76] Dursun G, Cicek HC, Dursun AY. Adsorption of phenol from aqueous solution by using carbonised beet pulp. Journal of Hazardous Materials B 2005; 125: 175–182.

[77] Patterson JF. Industrial Wastewater Treatment Technology, Second ed., Butterworths, London, 1985.

[78] Tepe O, Dursun AY. Combined effects of external mass transfer and biodegradation rates on removal of phenol by immobilized Ralstonia eutropha in a packed bed reactor. Journal of Hazardous Materials 2008; 151: 9-16.

[79] Knoll G, Winter J. Anaerobic degradation of phenol in sewage sludge: benzoate formation from phenol and carbon dioxide in the presence of hydrogen. Applied Environmental Microbiology 1987; 25(4): 384–391.

[80] El-Naas MH, Al-Muhtaseb SA, Makhlouf S. Biodegradation of phenol by Pseudomonas putida immobilized in polyvinyl alcohol (PVA) gel. Journal of Hazardous Materials 2009; 164: 720–725.

[81] Carrera J, Martín-Hernández M, Suárez-Ojeda ME, Pérez J. Modelling the pH dependence of the kinetics of aerobic p-nitrophenol biodegradation. Journal of Hazardous Materials 2011; 186: 1947–1953.

[82] Ye J, Singh A, Ward O. Biodegradation of nitroaromatics and other nitrogen containing xenobiotics. World Journal Microbiology Biotechnology 2004; 20: 117–135.

[83] Abrahamsson K, Klick S. Degradation of Halogenated Phenols in Anoxic Marine Sediments. Marine Pollution Bulletin 1991; 22: 227-233.

[84] Hakulinen R, Woods S, Ferguson J, Benjamin M. The Role of Facultative Anaerobic Microorganisms in Anaerobic Biodegradation of Chlorophenols. Water Science & Technology 1985; 17: 289-301.

[85] Jain V, Bhattacharya SK, Uberoi V. Degradation of 2,4-Dichlorophenol in Methanogenic Systems. Environmental Technology 1994; 15: 577-584.

[86] Leuenberger C, Giger W, Coney R, Graydon JW, Molnar-Kubica E. Persistent Chemicals in Pulp Mill Effluents. Water Research 1985; 19: 885-894.

[87] Sierra-Alvarez R, Field JA, Kortekaas S, Lettinga G. Overview of the Anaerobic Toxicity caused by Organic Forest Industry Wastewater Pollutants. Water Science Technology 1994; 29: 353-363.

[88] Wood JM. Chlorinated Hydrocarbons: Oxidation in the Biosphere. Environmental Science & Technology 1982: 16: 291A-297A.

[89] Annachhatre AP, Gheewala SH. Biodegradation of Chlorinated Phenolic Compounds. Biotechnology Advances 1996; 14 (1): 35-56.

[90] Cheremisinoff NP. Biological Degradation of hazardous Waste. In: Biotechnology for Waste and Wastewater Treatment. Noyes Publications: Westwood, New Jersey, USA. 1996. p37-110.

[91] Smith MJ, Lethbrideg G, Burns RG. Bioavailability and biodegradation of polycyclic aromatic hydrocarbons in soils. FEMS Microbiology Letters 1997; 152: 141–147.

[92] Yuan SY, Wei SH, Chang BV. Biodegradation of polycyclic aromatic hydrocarbons by a mixed culture. Chemosphere 2002; 41: 1463–1468.

[93] Ulrici W. Contaminated soil areas, different countries and contaminants, monitoring of contaminants, In: Rehm HJ., Reed G., Puhler A., Stadler P. (Eds.) Environmental Processes II Soil Decontamination Biotechnology: A Multi Volume Comprehensive Treatise, In: J. Klein (Ed.), Second Ed., vol. 11b, Wiley–VCH,Weihheim, FRG, 2000. p5-42.

[94] Odokuma LO, Dickson AA, Bioremediation of a crude oil polluted tropical rain forest soil. Global Journal of Environmental Science 2003; 2: 29–40.

[95] Cheung KC, Zhang JY, Deng HH, Ou YK, Leung HM, Wu SC, Wong MH. Interaction of higher plant (jute), electrofused bacteria and mycorrhiza on anthracene biodegradation. Bioresource Technology 2008; 99: 2148–2155.

[96] Somtrakoon K, Suanjit S, Pokethitiyook P, Kruatrachue M, Lee H, Upatham S. Enhanced biodegradation of anthracene in acidic soil by inoculated Burkholderia sp. VUN10013. Current Microbiology 2008; 57: 102–107.

[97] Li X, Lin X, Li P, Liu W, Wang L, Ma F, Chukwuka KS. Biodegradation of the low concentration of polycyclic aromatic hydrocarbons in soil by microbial consortium during incubation. Journal of Hazardous Materials 2009; 172: 601–605.

[98] Santos EC, Rodrigo JS, Jacques Bento FM, Peralba MDCR, Selbach PA, Enilso LSS, Camargo FAO. Anthracene biodegradation and surface activity by an iron-stimulated Pseudomonas sp. Bioresource Technology 2008; 99: 2644–2649.

[99]   Zeinali M, Vossoughi M, Ardestani SK. Naphthalene metabolism in Nocardia otiti-discaviarum strain TSH1, a moderately thermophilic microorganism. Chemosphere 2008; 72: 905–909.

[100] Hwang G, Park SR, Lee CH, Ahn IS, Yoon YJ, Mhin BJ. Influence of naphthalene bio-degradation on the adhesion of Pseudomonas putida NCIB 9816-4 to a naphthalene-contaminated soil. Journal of Hazardous Materials 2009; 171: 491–493.

[101] Gennaro PD, Rescalli E, Galli E, Sello G, Bestetti G, Characterization of Rhodococcus opacus R7, a strain able to degrade naphthalene and o-xylene isolated from a polycy-clic aromatic hydrocarbon-contaminated soil. Research in Microbiology 2001; 152: 641–651.

[102] Calvo C, Toledo FL, González-López J. Surfactant activity of a naphthalene degrad-ing Bacillus pumilus strain isolated from oil sludge. Journal of Biotechnology 2004; 109: 255–262.

[103] Kelley I, Freeman JP, Evans FE, Cerniglia CE. Identification of metabolites from deg-radation of naphthalene by a Mycobacterium sp. Biodegradation 1990; 1: 283–290.

[104] Lin C, Gan L, Chen ZL. Biodegradation of naphthalene by strain Bacillus fusiformis (BFN). Journal of Hazardous Materials 2010; 182: 771–777.

[105] Staples CA, Peterson DR, Parkerton TF, Adams WJ. The environmental fate of phtha-late esters: a literature review. Chemosphere 1997; 35: 667–749.

[106] Nalli S, Cooper DG, Nicell JA. Biodegradation of plasticizers by Rhodococcus rho-dochrous. Biodegradation 2002; 13: 343–352.

[107] Nalli S, Cooper DG, Nicell JA. Metabolites from the biodegradation of di-ester plasti-cizers by Rhodococcus rhodochrous. Science of the Total Environment Journal 2006; 366: 286–294.

[108] Rahman M, Brazel CS. The plasticizer market: an assessment of traditional plasticiz-ers and research trends to meet new challenges. Progress in Polymer Science 2004; 29: 1223–1248.

[109] Tickner JA, Schettler T, Guidotti T, McCally M, Rossi M. Health risks posed by use of di-2-ethylhexyl phthalate (DEHP) in PVC medical devices: a critical review. Ameri-can Journal of Industrial Medicine 2001; 39: 100–111.

[110] Onorato TM, Brown PW, Morris P. Mono-(2-ethylhexyl)phthalate increase spermato-cyte mitochondrial peroxiredoxin 3 and cyclooxygenase 2. Journal of Andrology 2008; 29: 293–303.

[111] Horn O, Nalli S, Cooper DG, Nicell JA. Plasticizer metabolites in the environment. Water Research 2004; 38: 3693–3698.

[112] Nalli SS, Horn OJ, Grochowalski AR, Cooper DG, Nicell JA. Origin of 2- ethylhexanol as a VOC. Environmental Pollution Journal 2006; 140: 181–185.

[113] Barnabé S, Beauchesne I, Cooper DG, Nicell JA. Plasticizers and their degradation products in the process streams of a large urban physicochemical sewage treatment plant. Water Research 2008; 42: 153–162.

[114] Beauchesne I, Barnabé S, Cooper DG, Nicell JA. Plasticizers and related toxic degradation products in wastewater sludges. Water Science & Technology 2008; 57: 367–374.

[115] Deligio T. Phthalate Alternative Recognized by ECHA. 2009.

[116] http://www. plasticstoday.com/articles/phthalate-alternative-recognized-echa/ (accessed December 2012)

[117] Gartshore J, Cooper DG, Nicell JA. Biodegradation of plasticizers by Rhodotorula Rubra. Environmental Toxicology & Chemistry 2003; 22: 1244–1251.

[118] Pour AK, Cooper DG, Mamer OA, Maric M, Nicell JA. Mechanism of biodegradation of dibenzoate plasticizers. Chemosphere 2009; 77: 258–263.

[119] Firlotte N, Cooper DG, Maric M, Nicell JA. Characterization of 1,5-pentanediol dibenzoate as a potential green plasticizer for poly(vinyl chloride). Journal of Vinyl Additive Technology 2009; 15: 99–107.

[120] Pour AK, Mamer OA, Cooper DG, Maric M, Nicell JA. Metabolites from the biodegradation of 1,6-hexanediol dibenzoate, a potential green plasticizer, by Rhodococcus rhodochrous. Journal of Mass Spectrometry 2009; 44: 662–671.

[121] Pour AK, Roy R, Coopera DG, Maric M, Nicell JA. Biodegradation kinetics of dibenzoate plasticizers and their metabolites. Biochemical Engineering Journal 2013; 70: 35-45.

[122] http://ec.europa.eu/environment/waste/framework/index.htm (accessed December 2012)

[123] Shimao M. Biodegradation of plastics. Current Opinion in Biotechnology 2001; 12: 242–247.

[124] Chiellini E, Corti A, D'Antone S, Solaro R. Biodegradation of poly(vinylalcohol) based materials. Progress in Polymer Science 2003; 28: 963-1014.

[125] Sakai K, Hamada N, Watanabe Y. Studies on the poly(vinyl alcohol)-degrading enzyme. Part VI. Degradation mechanism of poly(vinyl alcohol) by successive reactions of secondary alcohol oxidase and β-diketone hydrolase from Pseudomonas sp. Agricultural & Biological Chemistry 1986; 50: 989-996.

[126] Suzuki T. Degradation of poly(vinyl alcohol) by microorganisms. Journal of Applied Polymer Science Applied Polymer Symposium 1979; 35: 431-437.

[127] Hatanaka T, Kawahara T, Asahi N, Tsuji M. Effects of the structure of poly(vinyl alcohol) on the dehydrogenation reaction by poly(vinyl alcohol) dehydrogenase from Pseudomonas sp. 113P3. Bioscience Biotechnology Biochemistry 1995; 59: 1229-1231.

[128] Bloembergen S, David J, Geyer D, Gustafson A, Snook J, Narayan R. Biodegradation and composting studies of polymeric materials. In: Doi Y, Fukuda K. (Eds.) Biodegradable plastics and polymers. Amsterdam: Elsevier; 1994. p601-609.

[129] David C, De Kesel C, Lefebvre F, Weiland M. The biodegradation of polymers: recent results. Angewandte Makromolekulare Chemie 1994; 216: 21-35.

[130] Chiellini E, Corti A, Sarto GD, D'Antone S. Oxo-biodegradable polymers e Effect of hydrolysis degree on biodegradation behaviour of poly(vinyl alcohol). Polymer Degradation and Stability 2006; 91: 3397-3406.

[131] Khanna S, Srivastava AK, Recent Advances in microbial polyhydroxyalkanoates. Process Biochemistry 2005; 40: 607-619.

[132] Ghatnekar MS, Pai JS, Ganesh M. Production and recovery of poly-3-hydroxybutyrate from Methylobacterium sp.V49. Journal of Chemical Technology and Biotechnology 2002; 77: 444-448.

[133] DeMarco S. Advances in polyhydroxyalkanoate production in bacteria for biodegradable plastics. MMG 445. Basic Biotechnology eJournal 2005; 1: 1-4.

[134] Mudgal S, Lyons L, Bain, J. Plastic Waste in the Environment – Final Report for European Commission DG Environment. BioIntelligence Service; 2010. http://www.ec.europa.eu/environment/ (accessed December 2012)

[135] Hopewell J, Dvorak R, Kosior E. Plastics recycling: challenges and opportunities. Philosophical Transactions of the Royal Society B 2009; 364: 2115-2126.

[136] Plastics Europe. An analysis of European Plastics production, demand and recovery for 2009. Plastics - the Facts 2010.

[137] http://www.plasticseurope.org/ (accessed December 2012)

[138] The Encyclopedia of Polymer Science and Technology, 4th Edition, John Wiley and Sons, New York; 2012.

[139] Selke SE. Plastics recycling In: Harper CA. (Ed.), Handbook of plastics, elastomers and composites, 4th edition, McGraw-Hill, New York; 2002. p693–757.

[140] Fact sheet, Recycling the hard stuff. U.S. Environmental Protection Agency, Solid Waste and Emergency Response, 2002 EPA 530-F-02-023 Washington, D.C. http://www.docstoc.com/docs (accessed December 2012)

[141] Hendrickson CT, Matthews DH, Ashe M, Jaramillo P, McMichael FC. Reducing environmental burdens of solid-state lighting through end-of-life design. Environmental Research Letters 5. 2010. Doi: 10.1088/1748-9326/5/1/014016.

[142] Cerdan C, Gazulla C, Raugei M, Martinez E, Fullana-i-Palmer P. Proposal for new quantitative eco-design indicators: a first case study. Journal of Cleaner Production 2009; 17: 1638-1643.

[143] Plastic Waste: Redesign and Biodegradability. Science for Environmental Policy, Future Brief, 2001; 1: 1-8.

[144] Tokiwa Y, Calabia BP, Ugwu CU, Aiba S. Biodegradability of Plastics. International Journal of Molecular Science 2009; 10: 3722–3742.

[145] Jackson S, Bertényi T. Recycling of Plastics. ImpEE Project. 2006. p1-27

[146] http://www-g.eng.cam.ac.uk/ (accessed December 2012)

Chapter 2

Biodegradation of Nitrogen in a Commercial Recirculating Aquaculture Facility

S. Sandu and E. Hallerman

Additional information is available at the end of the chapter

1. Introduction

1.1. Need for biodegradation of nitrogen species in aquaculture systems

Commercial production of fish involves high levels of feeding. While digestive breakdown of lipids and carbohydrates yields water and carbon dioxide as waste products, digestion of proteins also yields nitrogenous compounds. In teleost (i.e., bony) fishes, these nitrogenous wastes are excreted predominately as ammonia. Total ammonia-nitrogen (TAN) consists of ionized ammonia (NH_4^+-N) and un-ionized ammonia (NH_3-N), the latter of which can prove toxic to fish. The fraction of TAN in the unionized form is dependent upon the pH and temperature of the water (Losordo 1997, Lekang 2007) and to a lesser degree its salinity (Diaz et al. 2012). At pH values less than 7.5, most ammonia is in the ionized form, and high levels of TAN can be tolerated. At higher pH, however, levels of un-ionized ammonia become problematic. Hence, biodegradation of ammonia is critical for the success of fish culture. Nitrifying bacteria, including *Nitrosomonas* sp., utilize NH_3-N as the energy source for growth, producing nitrite, NO_2-N. While nitrite-nitrogen is not as toxic as un-ionized ammonia-nitrogen, it can prove harmful to fish. The most common mode of toxicity is anoxia, as nitrite-nitrogen crosses the gills into the circulatory system and converts hemoglobin to methemoglobin, rendering it unable to bind and transport oxygen to the tissues (Palachek and Tomasso 1984, Svobodova et al. 2005). Other nitrifying bacteria, including *Nitrobacter* sp., utilize nitrite as their energy source, producing nitrate, NO_3-N. Nitrate-nitrogen concentrations are not generally of concern to aquaculturists, as most species can tolerate levels as high as 200 mg/L (Russo and Thurston 1991). Nitrate rarely reaches such high levels, as it is removed from the system by water exchanges and by passive denitrification in anaerobic pockets within the production or filtration systems (van Rijn 1996, Tal et al. 2006) or in denitrification reactors (Hamlin et al. 2008, Sandu et al. 2011).

Controlled degradation of nitrogenous wastes in filtration units is a major consideration in design and operation of commercial recirculating aquaculture systems. Among the technologies available (Crab et al. 2007), biological filtration is most commonly used. Biological filters are designed to provide abundant surface area for the attachment of complex microbial communities (Schreier et al. 2010) rich in *Nitrosomonas* and *Nitrobacter* species (Chen et al. 2006, Itoi et al. 2007, van Kessel et al. 2010). The nitrification capacity of the water treatment system is often the factor that limits production in a recirculating aquaculture system (Lemarie et al. 2004, Eschar et al. 2006, Diaz et al. 2012).

1.2. Utility of a nitrogen budget

The production efficiency of an aquaculture system can be evaluated through analysis of the conversion of nitrogen to fish biomass and to biodegradation pathways (Thoman et al. 2001). Nitrogen dynamics can be quantified by a mass balance equation, most simply as the difference between nitrogen in the feed supply and nutrients subsequently fixed as fish biomass. A nitrogen budget can quantify nitrogen fixation in fish biomass at various fish stocking densities (Suresh and Kwei 1992; Siddiqui and Al-Harbi 1999), nutrient release into the water column as dissolved and particulate excretion of fish (Krom and Neori 1989), and deposition of nitrogen into pond sediment (Acosta-Nassar et al. 1994). By estimating total nitrogen budgets for a particular species and culture system, we can evaluate the efficacy of water treatment processes (Porter et al. 1987). Hence, a nitrogen budget provides information crucial for the design and optimization of a production system, feeding strategies, and water and effluent treatment processes.

1.3. Biodegradation of nitrogenous wastes in a tilapia production system

Blue Ridge Aquaculture (BRA) in Martinsville, Virginia, USA is a large commercial facility that produces 1300 metric tons of hybrid tilapia *Oreochromis sp.* per year. To our knowledge, it is the largest recirculating aquaculture facility in existence. Before our study, little information existed about nitrogen budgets in commercial-scale fish production facilities, especially those using freshwater recirculating systems. By deriving a nitrogen budget, we can quantify the forms and proportions of nitrogen ingested as food as it becomes bound in tilapia biomass, excreted as metabolites, biodegraded by microorganisms, lost as gas by denitrification, or released in effluent. Knowledge of the nitrogen budget can help optimize operations, improving facility efficiency and maximizing production. Using Blue Ridge Aquaculture as our study system, our objectives were to: (1) examine nitrogen dynamics for the grow-out systems, (2) relate the nitrogen budget to water quality, (3) evaluate biofilter loading and nitrogen removal efficiency, and (4) predict maximum system carrying capacity. All abbreviations used in this chapter are shown in Table 1.

a = mole fraction of unionized ammonia nitrogen (decimal fraction)
ACR = areal conversion rate (mg/m²-d)
A_{NH3-N} = concentration of unionized ammonia nitrogen (mg/L)

A_{TAN} = maximum allowable concentration of total ammonia nitrogen (mg/L)

BRA = Blue Ridge Aquaculture

C_{TAN} = total ammonia nitrogen concentration in fish tank (mg/L)

C_{TANe} = total ammonia nitrogen concentration in the effluent from filters (mg/L)

C_{TANi} = total ammonia nitrogen concentration in the supply water (mg/L)

E_a = efficiency of rotating biological contactor for removal of ammonia nitrogen (percent)

FA = amount of feed (kg)

FB = fish biomass (kg)

FC = feed conversion factor (decimal fraction)

FP = protein content of feed (decimal fraction)

FR_{MTAN} = maximum feeding rate (kg/d)

$LC50$ = lethal concentration of a compound to 50% of the individuals in a population

LN = nitrogen load (g N/kg fish produced)

L_{TAN} = ammonia loading (g/hr)

N_{DENIT} = nitrogen gas removed by denitrification (mg/L)

N_{DIN} = dissolved inorganic nitrogen (mg/L)

N_{feed} = nitrogen fixed in feed (g/kg feed)

N_{fish} = nitrogen fixed in fish (g/kg fish produced)

N_{mort} = nitrogen fixed in dead fish (g/kg fish removed)

N_{NO2^-} = nitrite nitrogen (mg/L)

N_{NO3^-} = nitrate nitrogen (mg/L)

N_{NH3vol} = nitrogen removed by ammonia volatilization (mg/L)

NO_3-N_{pass} = nitrate removed passively by denitrification (mg/L)

NO_3-N_{exch} = nitrate removed by exchange of water (mg/L)

N_{TAN} = total ammonia nitrogen (mg/L)

N_{TON} = total organic nitrogen (mg/L)

PC = protein content of feed (decimal fraction)

P_{NO3-N} = partitioning of nitrate nitrogen (g/kg)

P_{TAN} = production rate of ammonia nitrogen (g/kg)

Q = flow rate through system (m³/min or L/h)

Q_f = recirculation flow rate (m³/min or L/h)

RAS = recirculating aquaculture system

R_{TAN} = ammonia removal rate (g/h)

S = surface area (m²)

SBM_{MTAN} = maximum biomass that could be sustained by system (kg fish)	
$TAN_{exchange}$ = ammonia removed by water exchange (mg/L)	
$TAN_{pass+vol}$ = ammonia removed by passive nitrification and ammonia volatilization (mg/L)	
$TAN_{RBC\ nitrification}$ = ammonia removed by nitrification in rotating biological contactor (mass/volume)	
t = time	
TKN = total Kjeldall nitrogen (g)	
TNI = total nitrogen input (kg/day)	
TNR = total nitrogen recovered (kg/day)	
$TNUA$ = total nitrogen unaccounted for (kg/day)	

Table 1. Abbreviations and associated units.

Tilapias are a group of fishes of great importance to world aquaculture (Costa-Pierce and Rakocy 1997, Fitzsimmons 1997, Lim and Webster 2006). Tilapias adapt readily to a range of production systems ranging from traditional extensive pond systems to high-input intensive pond systems to super-intensive recirculating aquaculture systems. Like all fishes, tilapias are sensitive to concentrations of nitrogenous wastes. The 48-hour LC_{50} value for NH_3 for Jordan tilapia *Oreochromis aureus* was 2.40 mg/L (Redner and Stickney 1979). The 48-hour LC_{50} value for hybrid red tilapia *O. mossambicus x O. niloticus* fry was 6.6 mg/L (Daud et al. 1988), although the threshold lethal concentration was 0.24 mg/L. The 24-hour LC_{50} value for un-ionized ammonia for *O. niloticus* was 1.46 mg/L (Evans et al. 2006) Sublethal effects of NH_3-N include tissue damage, decreased growth, increased feed conversion ratio, acute stress response, increased disease susceptibility, and reduced reproductive capacity (Russo and Thurston 1991, Yildiz et al. 2006, El-Sherif and El-Feky 2008, Benli et al. 2008). Tilapias also exhibit sensitivity to elevated nitrite concentrations. The 96-hour LC_{50} for nitrite-nitrogen for *O. aureus* was 16.2 mg/L at pH 7.2 and 22 mg/L chloride (Palachek and Tomasso 1984). Acute nitrite toxicity for *O. niloticus* varied with chloride levels and with fish size, with smaller fish proving more tolerant (Atwood et al. 2001, Wang et al. 2006). Nitrite-nitrogen levels should be kept below 5 mg/L within tilapia culture vessels (Losordo 1997). Knowledge of these toxicity values is useful for setting criteria for the design or evaluating the performance of filters for biodegradation of nitrogenous wastes in aquaculture systems.

2. Methods

2.1. Culture systems

The BRA facility includes systems for broodstock holding, fish breeding, egg incubation/ hatching, fingerling rearing, and food-fish production. The main building houses 42 recirculating aquaculture systems for grow-out to market size (Figure 1) that were the focus of our study. Each grow-out system (Figure 2) includes a fish production tank, a sedimentation basin

for solids removal, a rotating biological contactor (RBC) for microbial biodegradation including nitrification, and an oxygenation unit. Each fish production system is rectangular in shape,
built from concrete, holds 215 m³ of water, and consists of a fish-rearing tank (119 m³), a multi-
tube clarifier sedimentation basin (37 m³), an air-driven rotating biological contactor (59 m³
basin volume, 13,366 m² surface area per shaft), and an underground U-tube oxygenation
system. The total volume of the grow-out unit is 9030 m³. The water surface is at the same level
in the fish tank, sedimentation and RBC compartments, and water passes freely from one
section into another through large pipes or apertures. A pump receives water from the rotating
biological contactor compartment and pushes it through U-tubes and then to the far end of the
fish production tank, driving the recirculation. The filtration rate is 3.8 m³/min, and the system
turnover time is about once per hour.

Figure 1. Commercial-scale tilapia grow-out systems at Blue Ridge Aquaculture. The grow-out units are to the right of
the catwalk and sedimentation basins to the left. Photograph courtesy of Blue Ridge Aquaculture.

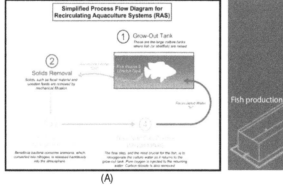
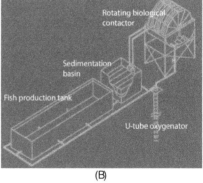

(A)   (B)

Figure 2. A) Conceptual diagram and (B) and engineering drawing of a single recirculating tilapia grow-out system at
Blue Ridge Aquaculture. Diagram courtesy of Blue Ridge Aquaculture.

BRA practices partial water exchange daily for controlling solids, dissolved organics and
nutrient accumulation in fish grow-out tanks. Water is exchanged daily from the system in the

interval between 2:00 p.m. and 8:00 a.m. Management practice is to completely flush the sedimentation basin after each instance that 227 kilograms of feed has been administered to a particular production unit. The exchange rate averages 22.3% per day, but the daily percentage varies among production units as a function of the size of fish, water quality requirements, and the amount of feed delivered to the system. The exchange water originates from wells, and is supplemented with municipal tap water when necessary. Exchange water replaces that used to remove settled particulate material, and thereby dilutes dissolved organic materials, dissolved nutrients, and salts.

Fish are fed commercially-prepared pelleted diets containing 36 or 40% minimum crude protein and 8-16% lipid levels, varying with the age of the fish. The feed is distributed hourly to the tanks over the 24-hour period. Fish production is managed so that 21-27 metric tons of 600-g fish reaches marketable size each week for shipment to a live market.

2.2. System boundaries

For the purpose of this study, the 42 recirculating aquaculture systems for grow-out were delimited as a unique system for purposes of quantifying the nitrogen budget. In certain contexts as set out below, N dynamics were quantified in greater detail in four individual systems. Broodstock holding and spawning facilities, a hatchery, and two greenhouses for fingerling production contain only a small part of the facility fish biomass, volume and exchange flow (i.e., they handle 3.0% of the fish biomass and 4.4% of the total nitrogen input). Because of their small contribution, the nitrogen budgets for these systems are not presented here, but can be found in Sandu (2004).

2.3. Inputs, outputs and nitrogen pools

The nitrogen budget is expressed as a mass-balance equation of all nitrogen forms, with total inputs plus generation equal to total outputs plus consumption. We found no measurable amounts of dissolved inorganic nitrogen in the replacement water. Hence, feed provided to fish was the sole nitrogen source in the form of organic nitrogen (N_{feed}). Multiplication of N_{feed} by the total amount of feed provided the mass of total nitrogen input (*TNI*). The nitrogen budget was accounted for in five known pools:

1. Nitrogen fixed in fish biomass as organic nitrogen, N_{fish},

2. Nitrogen fixed in dead fish biomass as organic nitrogen, N_{mort},

3. Dissolved inorganic nitrogen, N_{DIN}, which included N_{TAN}, N_{NO2}, and N_{NO3},

4. Total organic nitrogen in effluent, N_{TON}, and

5. Nitrogen gas removed from the system by passive denitrification, N_{denit}, and by ammonia volatilization, $N_{NH3\ vol}$.

All transformations among pools were assumed to be in a dynamic equilibrium over a defined period of time. We accounted for the mass fractions of nitrogen from Pools 1 to 4 (i.e., the measurable pools) as total nitrogen recovered (*TNR*), while the difference between total nitrogen input and total nitrogen recovered constituted pool 5, the mass fraction of total nitrogen unaccounted for (*TNUA*).

2.4. Analytical techniques

Analyses of fish and of feed for protein content followed Thiex et al. (2002), who indicated that by dry weight, 16% of protein is nitrogen. Samples were processed at the Forage Testing Laboratory, Virginia Polytechnic Institute and State University, Blacksburg, Virginia. Analyses for inorganic dissolved nitrogen forms (TAN, NO_2^--N, and NO_3^--N) were conducted on site using a Hach DR2400 spectrophotometer (Hach Company, Loveland, Colorado). Total Kjeldall nitrogen (TKN) was determined using macro-Kjeldall Standard Method 4500 – N_{org} B (APHA et al., 1998). Samples were acidified below pH 2 using H_2SO_4, refrigerated with ice, and transported to the Department of Civil and Environmental Engineering at Virginia Polytechnic Institute and State University, Blacksburg, Virginia, for analysis. Temperature and pH were measured directly on site using an Acorn Meter (Kit Model pH 6, Oakton, Vernon Hills, Illinois). Alkalinity was determined on-site using the Hach Permachem® Method. Dissolved oxygen (DO) was measured using a YSI (Model 550, Yellow Springs, Ohio) instrument. We calculated total organic nitrogen as the difference between TKN and total ammonia nitrogen (TAN).

2.5. Nitrogen budget determination

Under steady-state conditions, fish biomass does not fluctuate significantly over time (i.e., harvest equals growth), and the daily rations of feed are constant. Under these assumptions, we derived the nitrogen budget by determining the nitrogen input with feed and the output of nitrogenous compounds in known pools. We quantified daily amounts of nitrogen in feed, fish, and mortalities using information on feed consumption, fish production, and mortalities provided by BRA management. We measured the components of dissolved inorganic nitrogen and total organic nitrogen pools directly. We extrapolated mean values to the entire exchange volume from a day to determine the mass of nitrogen recovered in these forms. We assumed that the amount of nitrogen missing from the balance was lost by passive denitrification and by ammonia volatilization.

We considered both types of feed used in the system (with 36% or 40% standard protein content) to determine nitrogen fixed in feed, N_{feed}. We collected samples from three different points in storage silos for nitrogen content determination. We calculated N_{feed} as a composite using the equation:

$$N_{feed} = \Sigma \left(FA \times PC \times 0.16 \right) \tag{1}$$

where FA = amount of feed, PC = protein content of the feed, and 0.16 = concentration of nitrogen in protein (Thiex et al. 2002). We determined PC by laboratory analyses because protein content may differ from that claimed by the feed producer. We obtained the total mass of nitrogen originating from the feed input, *TNI*, by multiplying N_{feed} by the amount fed, *FA*.

To determine fixation of nitrogen in fish, N_{fish}, we analyzed protein content in triplicate samples of muscle tissue from fish from three size-classes. We estimated the proportions of fish in each size-class as 5% juveniles (i.e., newly introduced to the system from the hatchery), 60% intermediate, and 35% marketable size. With data on protein content of each fish size-class, we determined N_{fish} as a composite using the equation:

$$N_{fish} = \Sigma\left(FBxFPx0.16\right) \tag{2}$$

where FB = biomass of fish, and FP = protein content of the fish.

About 3.5% of the fish production (by number) was lost as mortalities. We assumed that nitrogen fixed in dead fish, N_{mort}, had the same nitrogen content as N_{fish}. In order to determine the biomass of N_{mort}, we collected mortalities daily from the production system for a two-week period, sorted them by size, and weighed them. We used these data to determine N_{mort} using equation 2.

Nitrogen load, L_N, entered the water column as ammonia and as organic nitrogen bound in feces. We quantified L_N as all nitrogen from feed that was not accounted for as living or dead fish as using the equation:

$$L_N = \left[N_{feed} - \left(N_{fish} + N_{mort}\right)\right] / FB \tag{3}$$

Hence, L_N quantified the amount of nitrogen that sustained the nitrogen cycle throughout the system, supplying all effluent nitrogen pools.

We quantified total organic nitrogen as the difference between TKN and TAN from the effluent. We obtained values for TKN, TAN, NO_2^--N, and NO_3-N by analyzing seven samples collected from the effluent discharge pipe at 3-hour intervals between 2:00 p.m. and 8:00 a.m. because effluent originated from the production system only during that interval. We repeated the tests twice (on different days) and averaged the results. We estimated daily production of these nitrogen forms by multiplying the average concentration (mg/L) by the volume of wastewater released from the system during a one-day period.

All nitrogen in feed that was not recovered as living or dead fish or as total organic nitrogen represented the dissolved inorganic fraction that entered the water as TAN. Hence, we determined ammonia production as:

$$P_{TAN} = N_{feed} - \left(N_{fish} + N_{mort} + N_{TON}\right) / FA \tag{4}$$

The sum of TAN, NO_2^-N, and NO_3^--N found in the effluent represented the fraction of nitrogen recovered as dissolved inorganic nitrogen, N_{DIN}. The summation of N_{DIN}, N_{fish}, N_{mort}, and N_{TON} provided the value for total nitrogen recovered, TNR. We determined total nitrogen unaccounted for, *TNUA*, by subtracting total nitrogen recovered, *TNR*, from total nitrogen input, *TNI*.

2.6. RAS carrying capacity, RBC design, TAN and NO$_3^-$-N removal

We used a simplified version of a model proposed by Losordo and Westers (1994) to determine the carrying capacity of the production system; that is, we considered only the parts of the model concerning maximum system carrying capacity with respect to TAN. Modeling of the flow rate through biofilters was unnecessary because the flow rate was fixed among all recirculating aquaculture systems at 3.78 m^3/min.

Four recirculating aquaculture systems chosen for intensive study held different age-groups of fish from juvenile to marketable size in order to represent the overall population in the facility. We knew total fish biomass, fish size, feeding rate, crude protein content of feed, daily percent body weight fed, flow rate through the system, and daily rate of exchange for each selected system. We measured other parameters, such as TAN, NO$_3^-$-N, NO$_2^-$-N, pH, temperature, and dissolved oxygen, using standard methods (APHA et al., 1998). We performed these analyses on composite samples collected from the fish-rearing tanks or from the rotating biological contactor's influent and effluent at four-hour intervals. By sampling from appropriate locations, we determined the effects of fish tanks, biofilters or sedimentation basins on each parameter. The experiments extended between consecutive water exchanges. We scaled the data to 24-hour intervals and determined mean and variance for each water quality parameter.

We determined the maximum system carrying capacity with respect to TAN as follows. We calculated the maximum allowable TAN concentration, A_{TAN}, as:

$$A_{TAN} = A_{NH3-N} / a \qquad (5)$$

where a = the mole fraction of unionized ammonia nitrogen as determined by pH and temperature (Huguenin and Colt, 1989). We calculated maximum feed rate, FR_{mTAN}, by assuming that the TAN concentration of a fish tank equals A_{TAN}, as:

$$FR_{mTAN} = \left[A_{TAN} x Q_f x E_a + Q\left(C_{TAN} - C_{TANi}\right)\right] / \left(0.092 x PC\right) \qquad (6)$$

where Q_f = the recirculating flow rate, or flow rate to the RBC, known to be 227,100 L/hr, and 0.092 = model constant coefficient. We determined the efficiency of the rotating biological contactor for removal of ammonia nitrogen, E_a as:

$$E_a = \left[\left(C_{TAN} - C_{TANe}\right)/C_{TAN}\right] \times 100 \qquad (7)$$

We estimated the maximum biomass that could be sustained within the system, SBM_{mTAN} as:

$$SBM_{mTAN} = FR_{mTAN} / \%BW \qquad (8)$$

where $\%BW$ = the feeding rate, expressed as a percent of body weight per day.

P_{TAN} is the rate of production of TAN in the system by metabolism of fish and microbial degradation of uneaten feed. We estimated P_{TAN} as a function of the feed rate and the percentage of protein in feed:

$$P_{TAN} = \left(FA * PC * 0.102\right) / t \qquad (9)$$

where t = the period of time from the onset of feeding to the next feeding.

This equation is based on the following assumptions and empirical estimates:

a.   16% of feed protein is nitrogen,

b.   80% of the nitrogen is assimilated,

c.   unassimilated nitrogen in fecal matter is removed rapidly from the tank,

d.   80% of assimilated nitrogen is excreted, and

e.   all of the TAN is excreted during t hours.

The coefficient 0.102 represents the product of values suggested by assumptions a through d (i.e., 0.16 x 0.8 x 0.8 = 0.102).

We determined the mass flow rate of TAN to a rotating biological contactor, or ammonia loading, L_{TAN}, from known (Q_t) and experimentally determined (C_{TANf}) parameters as:

$$L_{TAN} = Q_f x C_{TANf} \qquad (10)$$

We determined the ammonia removal rate, R_{TAN}, as:

$$R_{TAN} = \left(C_{TANf} - C_{TANe}\right) x Q_f \qquad (11)$$

The fraction $(R_{TAN} \times 100) / P_{TAN}$ represents the percentage of TAN that was removed by means other than the rotating biological contactor.

We estimated the nitrification performance of a rotating biological contactor as areal conversion rate, ACR, representing the amount of TAN oxidized by a unit of surface area in 24 hours:

$$ACR = R_{TAN} / S \qquad (12)$$

where S, the surface area of an RBC, was 13,336 m^2.

The mass balance quantifying the partitioning of P_{TAN} removal was:

$$P_{TAN} = TAN_{pass + vol} + TAN_{RBC\ nitrification.} + TAN_{exchange} \qquad (13)$$

We used a similar approach to determine NO$_3$-N partitioning using the equation:

$$P_{NO3\ -N} = NO_3^- - N_{pass.} + NO_3^- - N_{exch} \qquad (14)$$

2.7. Statistical analysis

We used linear regressions to determine the relationship between daily TAN production (P_{TAN}) and TAN removal efficiency per pass (E_a), and between fish biomass and percent P_{TAN} transformed by passive denitrification in the four systems tested.

3. Results

3.1. Nitrogen budget

We derived the nitrogen budget for the entire production system for mean conditions of 28.4ºC, pH 7.14, and alkalinity 119.0 mg/L as CaCO$_3$. For annual production of 1300 metric tons of fish biomass, BRA administers 2210 metric tons of feeds. These amounts correspond to 6054.8 kg feed consumed per day and 3561.6 kg fish weight gain per day. Of the feed utilized, 95% (5752.0 kg) was nominally 36% protein and 5% (302.8 kg) 40% protein content. However, laboratory analyses showed that the actual protein contents of the two feeds were somewhat lower, 35.0±0.2% and 39.8±0.2%, respectively. The estimated percentages of feed types and the laboratory-determined protein concentrations were used to determine the nitrogen fixed in feed, N_{feed} = 56.38 g/kg feed. By extrapolating N_{feed}, we determined a total nitrogen input of TNI = 341.381 kg/day.

Laboratory analyses showed that the three size-classes of fish from small to large had 18.04±0.16, 20.75±0.02 and 22.26±0.74% protein content, respectively. From these data, we determined that the nitrogen fixed in fish was N_{fish} = 33.83 g/kg produced. Extrapolating to the daily biomass of fish produced, the total nitrogen assimilated in fish was 120.49 kg/day.

Loss of fish represented 3.5% of the total production by number, with weighing of dead fish indicating losses of 2, 1, and 0.5% from the respective size-classes. This was the equivalent of 30.6 kg fish/day or 1.03 kg total N_{mort}/day, representing 0.86% of the total nitrogen assimilated. Hence, 35.3% of nitrogen from feed was assimilated in fish flesh (34.4% harvested and 0.86% removed with mortalities), and 64.7% was unassimilated or excreted in different forms. This latter term included nitrogen in uneaten feed that we accounted for in the overall budget as

N_{TON}. The nitrogen excreted, L_N, was 62.0 g/kg fish produced. Subsequently, the cumulative daily nitrogen loading for the entire system, L_N, was 221.3 kg.

Analyses of the effluent wastewater (estimated at 2017 m³/day) indicated, on average, 2.88 mg/L TAN, 1.09 mg/L NO_2^--N, 49.3 mg/L NO_3^--N, and 32.05 mg/L TON. Extrapolated to the entire effluent volume, the overall flows were 5.8 kg N_{TAN}/day (representing 1.70% of total nitrogen input, TNI), 2.2 kg N_{NO2}/day (0.64% TNI), 99.4 kg N_{NO3}/day (29.1% TNI), and 64.6 kg N_{TON}/day (18.9% TNI). Determination of total organic nitrogen, N_{TON}, allowed estimation of $P_{TAN} = 25.81$ g/kg feed. The recovered fraction of dissolved inorganic nitrogen, N_{DIN}, resulted from the summation:

$1.70\% N_{TAN} + 0.64\% N_{NO2} + 29.13\% N_{NO3} = 31.47\%$

Total nitrogen recovered, TNR, was determined as a percentage of TNI as:

$85.69\% TNR = 34.43\% N_{fish} + 0.86\% N_{mort} + 1.70\% N_{TAN} +$
$+ 0.64\% N_{NO2} + 29.13\% N_{NO3} + 18.93\% N_{TON}$

We then estimated total nitrogen unaccounted for, TNUA, as 14.3% of TNI. Hence, the subsequent nitrogen mass balance for the production system was:

341.381 kgTNI / day = 292.529 kgTNR / day + 48.852 kgTNUA / day

Table 2 summarizes the daily nitrogen budget for the production system. The relatively low value of total nitrogen unaccounted for, TNUA, was presumably due to nitrogen lost as nitrogen gas produced by denitrification and as ammonia lost to volatilization. Passive denitrification was likely the primary pathway because recirculated fish culture water passed through the sedimentation basin numerous times. As discussed below, the sediment blanket and associated thick biofilm in the multi-tube clarifier created anoxic conditions favorable for microbially mediated denitrification.

Units	Nitrogen pool							
	TNI	H_{fish}	N_{mort}	N_{TAN}	N_{NO2}	N_{NO3}	N_{TON}	TNUA
Kg	341.38	119.45	1.03	5.81	2.20	99.44	64.65	48.85
%	100.00	34.99	0.30	1.70	0.64	29.12	18.94	14.31

Table 2. Daily nitrogen budget for the grow-out system at Blue Ridge Aquaculture.

3.2. Carrying capacity, RBC design, TAN and NO_3^--N removal

The carrying capacity model indicated that recirculating aquaculture systems at Blue Ridge Aquaculture could support biodegradation of up to 3.15 mg TAN/L. This value corresponds to 0.025 mg/L maximum allowable unionized ammonia (A_{TAN}) at conditions of pH 7.0 and temperature of 30ºC (Huguenin and Colt 1989); our average values of these parameters for the four recirculating aquaculture systems monitored in greater detail were pH 7.09 and 27.8ºC. At 0.025 mg/L TAN, a recirculating system should be able to receive a maximum feeding rate of $FR_{max\ TAN}$ = 269.8 kg feed/day, which would support a fish biomass of $SBM_{max\ TAN}$ = 10,287.4 kg fish/system. Estimates of these parameters for each selected RAS are presented in Table 3. Comparison with actual feeding rates at the time of experiment (Table 4) showed that system loadings were 56.7 - 91.5% of the maxima estimated (Table 3, Figure 3). Over the four tanks examined in detail, TAN removal efficiency per pass, E_a, averaged 54.4%. We determined the rate of TAN production (P_{TAN}, Table 3). We determined P_{TAN} per kg of feed consumed by dividing these values by the daily amount of feed introduced into a system: i.e., 40.6 g/kg feed for feed with 40% crude protein content, and 36.7 g/kg feed for feed with 36% crude protein content. We found a positive, linear relationship between P_{TAN} (which also was proportional to the feeding rate) and E_a (slope = 0.0013, r^2 = 0.72), thereby showing that the rotating biological contactors efficiently removed various loadings of ammonia. None of the RBCs tested were working at maximum capacity.

Parameter	Units	RAS Tested				
		A12	A11	B16	A18	Average
Maximum feed rate (FR_{maxTAN})	kg/day	240.4	286.1	261.6	290.9	269.8
Maximum system biomass (SBM_{maxTAN})	kg	4202.5	11443.0	9871.0	15633.0	10287.4
Actual BW as % of SBM_{maxTAN}	%	56.66	91.52	66.54	76.77	72.87
TAN tank concentration	mg/L	1.77	2.32	2.04	2.10	2.06
TAN conc. in RBC influent ($CTAN_i$)	mg/L	1.77	2.32	2.04	2.10	2.06
TAN conc. in RBC effluent ($CTAN_e$)	mg/L	0.84	1.01	0.99	0.90	0.94
TAN removal efficiency per pass (E_a)	%	52.39	56.47	51.47	57.28	54.40
P_{TAN}/kg feed	g	40.6	36.7	36.7	36.7	37.7
Daily TAN production (P_{TAN})	g/day	5522.4	9626.4	6397.9	8161.9	7427.1
Ammonia loading (L_{TAN})	g/hr	402.19	526.87	463.28	478.50	467.71
Ammonia removal rate (R_{TAN})	g/hr	210.75	297.50	238.46	274.11	255.20
Areal conversion rate (ACR)	mg TAN/m²-d	378.4	534.2	428.2	429.2	442.5
ACR at SBW_{maxTAN}	mg TAN/m²-d	667.8	583.7	643.4	641.1	634
Mass TAN introduced by exchange	g/day	39.47	73.10	50.79	90.93	63.57
P_{TAN} introduced with water exchange	%/day	0.71	0.76	0.79	1.11	0.84
*Total TAN removed by water exchange	%/day	0.08	0.34	0.22	0.35	0.25

*Daily TAN percentage removal by water exchange, assuming that exchange water is treated using the treatment train tested by Sandu (2004) with 1.60 mg/L TAN.

Table 3. Experimentally determined and predicted parameters for estimation of maximum system carrying capacity with regard to TAN for tested units.

Parameter	Units	RAS Tested				
		A12	A11	B16	A18	Average
Water exchange rate	% volume/day	11.5	21.3	14.8	18.4	16.5
Flow rate through system (Q)	L/hr	1028.0	1903.7	1322.7	1645.8	1475.1
Fish size	g/fish	43	192	245	424	226
Fish biomass	kg	2381.0	10473.0	6568.5	12002.0	7856.1
Feeding rate (FR)	kg/day	136.0	262.0	174.0	222.3	198.6
Feed protein content (FP)	%	40	36	36	36	37
Percent body weight fed	kg feed/kg fish-d	5.72	2.50	2.65	1.85	3.18

Table 4. Characteristics of the recirculating aquaculture systems selected for evaluation.

Figure 3. Towards the end of a tilapia production cycle, stocking densities approach system carrying capacity. Photograph courtesy of Blue Ridge Aquaculture.

The mass flow-rate of TAN to a rotating biological contactor, L_{TAN}, averaged 467.7 g/hr, which was removed at an average rate of R_{TAN}= 255.2 g/hr. Per-system values are presented in Table 3. The ratio of R_{TAN} to P_{TAN} showed that rotating biological contactors removed an average of 84.0% of total ammonia nitrogen from the selected systems. From the difference, 1.1% of TAN was recovered from exchanged water and 15.0% remained unaccounted for, probably transformed to NO_2^--N and NO_3^--N by passive nitrification or lost by volatilization of ammonia. Data in Table 3 show that fish biomass in the system was positively correlated with the percentage of total ammonia nitrogen transformed by passive nitrification (slope = 0.0015, r^2 = 0.69); although the correlation was not strong, it shows that systems with higher biomass had lower water quality and larger microbial populations, including nitrifiers that promoted in-situ biodegradation of ammonia.

The rotating biological contactors removed between 378.4 and 534.2 mg TAN/m²/day (442.5 mg TAN/m²/day on average, Table 3). The areal conversion rate, ACR, increased with the loading of total ammonia nitrogen. Average ACR under conditions of maximum system biomass was estimated at 634.0 mg TAN/m²/day. We note that the difference between existing ACR and predicted maximum ACR is consistent with that between the existing fish biomass and predicted maximum fish biomass.

We derived a daily nitrogen budget partitioning the total ammonia nitrogen removal from each RAS (Table 5). On average among systems, 84.0% of TAN was removed by rotating biological contactors, 14.9% by passive nitrification and ammonia volatilization, and only 1.1% was removed by periodic water exchange.

System	[1]P_{TAN}		[2]$TAN_{pass + vol}$		[3]$TAN_{RBC\ nitrification}$		[4]$TAN_{exchange}$	
	g	%	g	%	g	%	g	%
A12	5522.4	100	421.30	7.63	5057.41	91.58	43.69	0.79
A11	9626.4	100	2380.50	24.73	7139.90	74.17	106.00	1.10
B16	6397.9	100	610.22	9.54	5722.92	89.45	64.76	1.01
A18	8161.9	100	1463.66	17.93	6578.49	80.60	119.75	1.47
Average	7427.2	100	1108.51	14.93	6235.09	83.95	83.55	1.12

[1] TAN production over a 24-hour period.

[2] TAN removed by passive nitrification and by ammonia volatilization.

[3] TAN removed by nitrification in RBC.

[4] TAN removed with exchanged water.

Table 5. Partitioning of total ammonia nitrogen removal for each recirculating aquaculture system studied.

We conducted tests on the same recirculating aquaculture systems to determine the fate of NO_3^--N following its production by nitrification. We regarded P_{NO3^--N} as approximately equal to P_{TAN} by assuming that TAN lost from the systems by water exchange and volatilization was negligible. Data on P_{NO3^--N}, water exchange rates, and NO_3^--N concentrations before and after water exchange allowed us to determine the total mass of NO_3^--N in the systems at these times and the amounts of NO_3^--N lost by water exchange and passive denitrification. That is, we derived a daily mass balance quantifying P_{NO3^--N} removal pathways from each RAS (Table 6). Results indicated that NO_3^--N accumulation was in the range of 9.1 – 17.2 mg/L in each RAS over a 24-hour period. On average, 44.1% of NO_3^--N was removed by water exchange, and the difference of 55.9% was removed by passive denitrification. NO_3^--N in effluent could be subject to microbial denitrification if water reuse is implemented (Sandu et al. 2008).

Parameter	Units	RAS Tested				
		A12	A11	B16	A18	Average
Daily NO_3^--N production ($P_{NO3^-\ -N}$)	g	5522.4	9626.4	6397.9	8161.9	7427.9
NO_3^--N conc. before exchange	mg/L	57.3	57.3	50.9	49.1	53.6
System mass NO_3^--N before exchange	g	12290.85	12290.85	10918.05	10531.95	11507.92
NO_3^--N conc. after exchange	mg/L	40.5	40.1	38.9	40.0	39.9
System mass NO_3^--N after exchange	g	8687.25	8601.45	8344.05	7872.15	8376.22
NO_3^--N and removed by exchange	g/day	3603.6	3689.4	2574.0	2659.8	3132.45

Parameter	Units	RAS Tested				
		A12	A11	B16	A18	Average
$P_{NO_3^- \text{-}N}$ removed by exchange	%/day	65.25	38.33	40.23	32.58	44.10
NO_3^--N lost by passive denitrification	g/day	1918.8	5937.0	3823.9	5502.1	4295.45
$P_{NO_3^- \text{-}N}$ lost by passive denitrification	%/day	34.75	61.67	59.77	67.42	55.90

Table 6. Dynamics and partitioning of $P_{NO_3^- \text{-}N}$ removal for each recirculating aquaculture system studied.

NO_2^--N always remained at concentrations lower than 0.3 mg/L in the fish tanks. Its concentration increased slightly as water passed through the sedimentation basin, but decreased again to concentrations lower that those in fish tanks after contact with the RBC, creating an equilibrium concentration. Because NO_2^--N concentrations were generally stable and below levels considered a threat to fish, we pursued no further determination of NO_2^--N dynamics.

4. Discussion

We quantified nitrogen fixation and biodegradation through the recirculating tilapia production system at Blue Ridge Aquaculture, a large commercial production facility. The 34.4% of total nitrogen input assimilated by the fish indicated excellent nitrogen utilization relative to other production systems. For example, Suresh and Kwei (1992) found that less than 20% of nitrogen was assimilated by tilapia using feed with 22% crude protein content and much lower fish stocking densities than those at BRA. Using feed with 34% crude protein content, Siddiqui and Al-Harbi (1999) reported 21.4% nitrogen assimilation by red tilapia. Although Suresh and Kwei (1992) found decreasing nitrogen assimilation with increasing fish density, Refstie (1977), Rakcocy and Allison (1981), and Vijayan and Leatherland (1988) reported the opposite finding. We attribute the high nitrogen assimilation in our study to higher protein content in feeds used at BRA, well-managed water quality, and to production of selectively bred fish (Hallerman 2000). Also, most earlier studies reported higher mortality rates, diminishing total nitrogen accumulated in fish.

The small amounts of nitrogen recovered as TAN and NO_2^--N likely were due to biodegradation in rotating biological contactors, which oxidized them effectively to NO_3^--N. Most of the nitrogen recovered as total organic nitrogen (18.93%) was probably due to feces, noting that feed was consumed by fish almost instantly at distribution, and that only fine particulates could escape as wasted feed. Assuming that some organic nitrogen in feces dissolved upon contact with water, our results with tilapia, which accounted for nitrogen from the entire organic pool, broadly agree with those of Porter et al. (1987, who found 10% fecal nitrogen) and Thoman et al. (2001, who recovered 14% nitrogen from suspended solids) for other species.

For total nitrogen unaccounted for (14.31%), removal of N_2 gas through passive denitrification is the most reasonable explanation. Although denitrification may seem surprising given the relatively high dissolved oxygen in the recirculating systems, development of anoxic microsites in sediment provides likely sites for denitrification (Brandes and Devol 1997). Anoxic microsites could arise in fish tanks where particles accumulated, or more likely, in the sedimentation basin, where a blanket of sediments developed for 19 – 36 hours before removal. We observed that large amounts of gases rapidly collected beneath the water surface in the sedimentation basin; however, samples we collected were contaminated with oxygen, precluding evaluation of biologically-generated nitrogen production. A thick biofilm on the tanks' walls also could have provided anoxic microsites, contributing to NO_3^--N removal. This explanation was supported by our results for Tank A12, where fish were harvested and the biofilm removed from the walls less than two weeks before our monitoring began. The time for regrowth of the biofilm to a thickness that could allow denitrification was limited. Subsequently, less than 35% of NO_3^--N production was removed by passive denitrification from this particular system, considerably less than in the other three systems monitored. In-situ denitrification has been reported by other authors. For example, Bovendeur et al. (1987) found that 40 – 80% of TAN oxidized by nitrification then was reduced by denitrification. Thoman et al. (2001) attributed 9 – 21% losses of systems' nitrogen to denitrification. The 56% removal of NO_3^--N by passive denitrification in our study represented an important, positive outcome, because it could reduce by more than half the investment necessary for nitrogen removal should the effluent be treated and reused as suggested by Sandu et al. (2008, 2011).

Our results indicated that despite high fish densities maintained at BRA, the systems are not being operated at their maximum carrying capacity. Our results showed that an average of 73% of the recirculating systems' productive potential was utilized, although utilization approached 92% in systems holding fish close to harvest size. In particular, much productive potential can be realized in systems holding smaller fish for long periods. By better distributing fish biomass among systems via more frequent grading, net production could be increased within existing space. Our suggestion for increased production is supported by the excellent average removal efficiency for rotating biological contactors (54.4%) at a recirculation rate of almost one pass per hour, and by an average areal conversion rate of 442.5 mg TAN/m^2/day, which maintained an average TAN of 2.06 mg/L in fish tanks. Up to 2830 mg TAN/m^2/day can be removed by a rotating biological contactor (Rogers and Klemeston 1985), suggesting that the biofilters could function successfully under the maximum conditions of 3.15 mg/L TAN and 634 mg TAN/m^2/day areal conversion rate that we predicted. Additionally, reusing water using a treatment train such as that described by Sandu (2004) and Sandu et al. (2008, 2011) with 1.6 mg/L TAN, only 0.84% of total ammonia nitrogen produced would be reintroduced to the recirculating systems. This additional loading would be removed easily by the rotating biological contactors, without significant increase of TAN throughout the systems.

5. Conclusion

Routine aquaculture production generates waste products for which controlled biodegradation in treatment units is a major consideration in design and operation of recirculating aquaculture systems. Biodegradation of nitrogenous wastes is critical, especially for unionized ammonia and nitrite, which are toxic to fish. We quantified the dynamics of nitrogen through a large commercial recirculating aquaculture facility producing hybrid tilapia *Oreochromis* sp. Our nitrogen budget evaluated total ammonia nitrogen (TAN) production and removal in biofilters, quantifying the fate of nitrate-nitrogen (NO_3^--N) and determining the systems' maximum carrying capacity under steady-state conditions. Most of the recovered nitrogen was in fish, nitrate-nitrogen, and total organic nitrogen pools, with relatively small proportions as total ammonia nitrogen, mortalities, and nitrite-nitrogen, totaling 86%. The remaining 14% of the nitrogen budget unaccounted for likely was lost by passive denitrification to nitrogen gas and by volatilization of ammonia. Our nitrogen biodegradation model predicts that the systems could operate safely at up to 3.15 mg/L total ammonia nitrogen. Under current production conditions, system loading was 57-92% of the maximum fish biomass that could be supported. The biofilters' areal conversion rate could be increased by half under conditions of maximum biomass loading. NO_2^--N was not a parameter of concern, always remaining below 0.3 mg /L. Our results showed that microbial biodegradation of fish wastes was more than adequate and that fish production could be increased within the existing farm infrastructure, especially by more frequent grading of fish in order to stock production systems at densities approaching carrying capacity. With denitrification, discharged culture water could be reused to realize savings in operating costs. Beyond the narrow interest in our study system, our approach can be applied more broadly to other fish culture systems.

Acknowledgements

We are grateful for the support of the Commercial Fish and Shellfish Technologies program and the Department of Fish and Wildlife Conservation at Virginia Tech University. Blue Ridge Aquaculture graciously allowed access to facilities and production records. Julie Petruska trained S.S. water quality testing procedures. The expertise of Nancy Love was indispensable in experimental design and analysis.

Author details

S. Sandu and E. Hallerman*

*Address all correspondence to: ehallerm@vt.edu

Department of Fish and Wildlife Conservation, Virginia Polytechnic Institute and State University, Blacksburg, VA, USA

References

[1] Acosta-Nassar, M.V.; Morrell, J.M., & Corredor, J.E. (1994). The nitrogen budget of a tropical semi-intensive freshwater fish culture pond. *Journal of the World Aquaculture Society*, Vol. 25, No. 2, pp. 261-270, ISSN 0893-8849

[2]   Atwood, H.L.; Fontenot, Q.C., Tomasso, J.R., & Isely, J.J. (2001). Toxicity of nitrite to
      Nile tilapia: effect of fish size and environmental chloride. *North American Journal of
      Aquaculture*, Vol. 63, No. 1, pp. 49-51, ISSN 1522-2055

[3]   APHA (American Public Health Association), American Water Works Association &
      Water Environment Federation (1998). *Standard Methods for the Examination of Water
      and Wastewater, 20th edition*. American Public Health Association, ISBN 0-875530235-7,
      Washington, DC

[4]   Benli, A.C.K.; Koksal, G., & Ozkul, A. (2008). Sublethal ammonia exposure of Nile ti-
      lapia (*Oreochromis niloticus* L.): effects on gill, liver, and kidney histology. *Chemo-
      sphere*, Vol. 72, No. 9, pp. 1355-1358, ISSN 0045-6535

[5]   Bovendeur, J.; Eding, E.H., & Henken, A.M. (1987). Design and performance of a wa-
      ter recirculation system for high-density culture of the African catfish, *Clarias gariepi-
      nus* (Burchell 1822). *Aquaculture*, Vol. 63, No. 1-4, pp. 329-353, ISSN 0044-8486

[6]   Brandes, J.A. & Devol, A.H. (1997). Isotopic fractionation of oxygen and nitrogen in
      coastal marine sediments. *Geochimical et Cosmochimical Acta*, Vol. 61, pp. 1798-1801,
      ISSN 0016-7037

[7]   Chen, S.; Ling, J., & Blancheton, J. (2007). Nitrification kinetics of biofilm as affected
      by water quality factors. *Aquacultural Engineering*, Vol. 34, No 3, pp.179-197, ISSN
      0144-8609

[8]   Colt, J. (2006). Water quality requirements for reuse systems. *Aquacultural Engineer-
      ing*, Vol. 34, No. 3, pp. 143-156, ISSN 0144-8609

[9]   Costa-Pierce, B.A. & Rakocy, J.E. eds. (1997). *Tilapia Aquaculture in the Americas*.
      World Aquaculture Society, Baton Rouge, LA. 522 pp. in two volumes, ISBN
      1-88807-01-6 and 1-88897-04-0

[10] Crab, R.; Avnimelech, Y., Defoird, T., Bossier, P., & Verstraete, W. (2007). Nitrogen
      removal techniques in aquaculture for sustainable production. *Aquaculture*, Vol. 270,
      No. 1-4, pp. 1-14, ISSN 0044-8486

[11] Daud, S.K.; Hasbollah, D., & Law, A.T. (1988). Effects of unionized ammonia on red
      tilapia (*Oreochromis mossambicus/O. niloticus* hybrid) fry. Pages 411-413 in Pullin,
      R.S.V., Bhukaswan, T., Tonguthai, K., and Maclean, J.L. *The Second International Sym-
      posium on Tilapia in Aquaculture*. ICLARM Conference Proceedings 15. Thailand De-
      partment of Fisheries and International Center for Living Aquatic Resources
      Management, ISBN 971-1022-58-3, Manila

[12] Diaz, V.; Ibanez, R., Gomez, P., Urtiaga, A.M., & Ortiz, I. (2012). Kinetics of nitrogen
      compounds in a commercial marine recirculating aquaculture system. *Aquacultural
      Engineering*, Vol. 50, No. 1, pp. 20-37, ISSN 0144-8609

[13] El-Sherif, M.S.; & El-Feky-A.M. (2008) Effect of ammonia on Nile tilapia (*O. niloticus*)
      performance and some hematological and histological measures. *8th International*

Symposium on Tilapia in Aquaculture, Cairo, Egypt, October 12-14, 2008, http://ag.arizo-na.edu/azaqua/ista/ISTA8/FinalPapers/PDF%20Files/39%20mohamed %20shreif12.pdf

[14] Eshchar, M.; Lahav, O., Mozes, N., Peduel, A, & Ron, B., (2006). Intensive fish culture at high ammonium and low pH. *Aquaculture,* Vol. 255, No. 1-4, pp. 301-313, ISSN 0044-8486

[15] Evans, J.J.; Pasnik, D.J., Brill, G.C., & Klesius, P.H. (2006) Un-ionized ammonia exposure in Nile tilapia: toxicity, stress response, and susceptibility to *Streptococcus agalactiae. North American Journal of Aquaculture,* Vol. 68, No. 1, pp. 23-33, ISSN 1522-2055

[16] Fitzsimmons, K., ed. (1997). *Tilapia aquaculture: Proceedings from the Fourth International Symposium on Tilapia in Aquaculture.* Natural Resource, Agriculture, and Engineering Service, Cooperative Extension, 152 Riley-Robb Hall, Ithaca, NY. 808 pp. in two volumes, ISBN 0-935817-58-1

[17] Hallerman, E.M. (2000). Genetic improvement of fishes for commercial recirculating aquaculture systems: a case study involving tilapia. In Libey, G.S. & Timmons, M.B., eds. Proceedings of the Third International Conference on Recirculating Aquaculture, Roanoke, VA, July 20-23, 2000.

[18] Hamlin, H.J.; Michaels, J.T., Beaulaton, C.M., Graham, W.F., Dutt, W., Steinbach, P., Losordo, T.M., Schrader, K.K., and Main, K.L. (2008). Comparing denitrification rates and carbon sources in commercial upflow denitrification biological filters in aquaculture. *Aquacultural Engineering,* Vol. 38, No. 2, pp. 79-92, ISSN 0144-8609

[19] Huguenin, J.E. & Colt, J. (1989). *Design and operating guide for aquaculture seawater systems.* Elsevier Interscience, ISBN 0-444-50577-6, Amsterdam. 264 pp

[20] Itoi, S.; Ebihara, N., Washio, S, and Sugita, H. (2007). Nitrate-oxidizing bacteria, *Nitrospira,* distribution in the outer layer of the biofilm from filter materials of a recirculating water system for the goldfish *Carassius auratus. Aquaculture,* Vol. 264, No. 1-4, pp. 297-308, ISSN 0044-8486

[21] Krom, M.D. & Neori, A. (1989). A total nitrogen budget for an experimental intensive fishpond with circularly moving seawater. *Aquaculture,* Vol. 83, No. 1-2, pp. 345-358, ISSN 0044-8486

[22] Krom, M.D.; Porter, C., & Gordin, H. (1985). Nutrient budget of a marine fish pond in Eilat, Israel. *Aquaculture,* Vol. 51, No. 1, pp 65-80, ISSN 0044-8486

[23] Lekang, O.I. (2007). Ammonia removal. Chapter 9 in *Aquaculture Engineering.* Blackwell Publishing, ISBN 978-1-4051-2610-6, Oxford

[24] Lemarie, G.; Dosdat, A., Coves, D., Dutto, G., Gasset, E., & Person-Le Ruyet, J. (2004). Effect of chronic ammonia exposure on growth of European seabass (*Dicentrarchus labrax*) juveniles. *Aquaculture,* Vol. 229, No. 1-4, pp. 479-491, ISSN 0044-8486

[25] Lim, C. & Webster, C.D., eds. (2006). *Tilapia Biology, Culture, and Nutrition*. The Haworth Press, Binghamton, New York, NY, USA, ISBN 13: 978-1-56022-318-4

[26] Losordo, T.M. (1997). Tilapia culture in intensive recirculating systems. Pages 185-211 in Costa-Pierce, B.A. & Rakocy, J.E. *Tilapia Aquaculture in the Americas, volume 1*. World Aquaculture Society, ISBN 1-88807-01-6, Baton Rouge, Louisiana, USA

[27] Losordo, T.M., and Westerman, P.W. (1994). An analysis of the biological, economic, and engineering factors affecting the cost of fish production in recirculating aquaculture systems. *Journal of the World Aquaculture Society* Vol. 25, No. 2, pp. 193-203. ISSN 1749-7345

[28] Losordo, T.M. & Westers, H. (1994). System carrying capacity and flow estimation. In: M.B. Timmons & T.M. Losordo, editors. *Aquaculture water reuse system: Engineering design and management*. Developments in Aquaculture and Fisheries Sciences, vol. 27. Elsevier Science, ISBN 9780444895851, Amsterdam. Pp. 9-60

[29] Palacheck, R.M. & Tomasso, J.R. (1984). Toxicity of nitrite to channel catfish (*Ictalurus punctatus*), tilapia (*Tilapia aurea*), and largemouth bass (*Micropterus salmoides*): evidence for a nitrite exclusion mechanism. *Canadian Journal of Fisheries and Aquatic Sciences* Vol. 41, No. 12, pp. 1739-1744, ISSN 1205-7533

[30] Porter, C.B.; Krom, M.D., Robbins, M.G., Brickell, L., & Davidson, A. (1987). Ammonia excretion and total N budget for gilthead seabream (*Sparus aurata*) and its effect on water quality conditions. *Aquaculture*, Vol. 66, No. 3-4, pp. 287-297, ISSN 0044-8486

[31] Refstie, T. (1977). Effects of density on growth and survival of rainbow trout. *Aquaculture*, Vol. 10, No. 3, pp. 231-242, ISSN 0044-8486

[32] Racocy, J. & Allison, R. (1981). Evaluation of a closed recirculating system for the culture of tilapia and aquatic macrophytes. In: Fish Culture Section, *Bioengineering Symposium for Fish Culture*, American Fisheries Society, Bethesda, Maryland, USA, Pages 296-307

[33] Redner, B.D. & Stickney, R.R. (1979). Acclimation to ammonia by *Tilapia aurea*. *Transactions of the American Fisheries Society* Vol. 108, No. 4, pp. 383-388, ISSN 1548-8659

[34] Rogers, G.L. & Klemetson, S.L. (1985). Ammonia removal in selected aquaculture water reuse biofilters. *Aquacultural Engineering*, Vol. 4, No. 2, pp. 135-154, ISSN 0144-8609

[35] Russo, R.C. & Thurston, R.V. (1991). Toxicity of ammonia, nitrite and nitrate to fishes. Pages 58-89 in D.E. Brune and J.R. Tomasso, eds. *Aquaculture and water quality*. Advances in World Aquaculture 3. World Aquaculture Society, Baton Rouge, Louisiana, USA.

[36] Sandu, S.I. (2004). Evaluation of ozone treatment, pilot-scale wastewater treatment plant, and nitrogen budget for Blue Ridge Aquaculture. Ph.D. dissertation, Virginia Polytechnic Institute and State University, Blacksburg, Virginia, USA.

[37] Sandu, S.; Brazil, B., & Hallerman, E. (2008). Efficacy of a pilot-scale wastewater treatment plant upon a commercial aquaculture effluent: I. Solids and carbonaceous compounds. *Aquacultural Engineering*, Vol. 39, No. 1, pp. 78-90, ISSN 0144-8609

[38] Sandu, S.; Brazil, B., & Hallerman, E. (2011). Efficacy of pilot-scale wastewater treatment upon a commercial recirculating aquaculture facility effluent. Pages 141-158 in B. Sladonja, ed. *Aquaculture and the Environment: A Shared Destiny*. Intech. ISBN 978-953-307-749-9, Rijeka, Croatia.

[39] Schreier, H.J.; Mirzoyan, N., & Saito, K. (2010). Microbial diversity of biological filters in recirculating aquaculture. *Current Opinion in Biotechnology*, Vol. 21, No. 3, pp. 318-325, ISSN 0958-1669

[40] Siddiqui, A.Q. & Al-Harbi, A.H. (1999). Nutrient budget in tanks with different stocking densities of hybrid tilapia. *Aquaculture*, Vol. 170, No. 3-4, pp. 245-252, ISSN 0044-8486

[41] Suresh, A.V. & Kwei, L.C. (1992). Effect of stocking density on water quality and production of red tilapia in a recirculating water system. *Aquacultural Engineering*, Vol. 11, No. 1, pp. 1-22, ISSN 0144-8609

[42] Svobodova, Z.; Machova, J., Poleszczuk, G., Huda, J., Hamackova, J., & Kroupova, H. (2005). Nitrite poisoning of fish in aquaculture facilities with water-recirculating systems. *Acta Veterinaria Brno*, Vol. 74, pp. 129-137, ISSN 0001-7213

[43] Tal, Y.; Watts, J.E.M., and Schreier, H.J. (2006). Anaerobic ammonium-oxidizing (anamox) bacteria in associated activity in fixed-film biofilters of a marine recirculating aquaculture system. *Applied and Environmental Microbiology*, Vol. 72, No. 4, pp. 2896-2904, ISSN 0099-2240

[44] Tchobanoglous, G. & Schoeder, E.D. (1985). *Water Quality: Characterization, Modeling, Modification*. Addison-Wesley Publishing Company. ISBN: 10-0201054337, Reading, MA

[45] Thiex, N.J.; Manson, H., Anderson, S., & Persson, J. A. (2002). Determination of crude protein in animal feed, forage, grain, and oilseeds by using block digestion with a copper catalyst and steam distillation into boric acid: collaborative study. *Journal of the Association of Official Agricultural Chemists*, Vol. 85, pp. 309-317, ISSN 0095-9111

[46] Thoman, E.S.; Ingall, E.D., Davis, D.A., & Arnold, C.R. (2001). A nitrogen budget for a closed, recirculating mariculture system. *Aquacultural Engineering*, Vol. 24, No. 3, pp. 195-211, ISSN 0144-8609

[47] van Kessel, M.A.J.H.; Harhangi, H.R., van de Pas-Schoonen, K., van de Vossenberg, J., Flik, G., Jetten, M.S.M., Klaren, P.H.M., & Op den Camp, H.J.M. (2010). Biodiversi-

ty of N-cycle bacteria in nitrogen removing bed biofilters for freshwater recirculating aquaculture systems. *Aquaculture*, Vol. 306, No. 1-4, pp. 177-184, ISSN 0044-8486

[48] van Rijn, J. (1996). The potential for integrated biological treatment systems in recirculating fish culture: A review. *Aquaculture* Vol. 139, No. 3-4, pp. 181-201, ISSN 0044-8486

[49] Vijayan, M.M. & Leatherland, J.F. (1988). Effects of stocking density on the growth and stress response in brook charr *Salvelinus fontinalis*. *Aquaculture*, Vol. 75, No. 1-2, pp. 159-170, ISSN 0044-8486

[50] Wang, Y.; Zhang, W., Li, W., & Xu, Z. (2006). Acute toxicity of nitrite on tilapia (*Oreochromis niloticus*) at different external chloride concentrations. *Fish Physiology and Biochemistry*, Vol. 32, No. 1, pp. 49-54, ISSN 0920-1742

[51] Yildiz, H.Y.; Koksal, G., Borazan, G., & Benli, C.K. (2006). Nitrite-induced methemoglobinemia in Nile tilapia, *Oreochromis niloticus*. *Journal of Applied Ichthyology*, Vol. 22, No. 5, pp. 427-426, ISSN 1439-0426

Biodegradation and Mechanical Integrity of Magnesium and Magnesium Alloys Suitable for Implants

S. González, E. Pellicer, S. Suriñach, M.D. Baró and
J. Sort

Additional information is available at the end of the chapter

1. Introduction

Most conventional orthopedic implants used for repairing joint and bone fractures consist of metallic biomaterials with polycrystalline microstructure that exhibit high hardness, good corrosion resistance and excellent fatigue and wear resistance. Usually, once the patient has recovered from a traumatic injury, a revision surgery is necessary in order to remove the implant from the body and avoid problems associated with osteopenia, inflammation of adjacent tissues or sarcoma. Alternatively, to avoid post-extraction of the implant, intensive efforts are being made in recent years to develop new classes of so-called "biodegradable implants", composed of non-toxic materials that become reabsorbed by the human body after a reasonable period of time. These implants are usually based on polymeric materials. However, polymeric implants are often rather costly and exhibit relatively low mechanical strength. Sometimes organic polymers can also react with human tissues, leading to osteolysis. For these reasons, it is highly desirable to develop cost-effective biodegradable metallic alloys, with better mechanical performance than polymers.

Although biodegradation is usually associated with the breakdown of organic matter into simple chemicals through the action of microorganisms, metals can also undergo biodegradation. Although corrosion should be generally avoided in the engineering field, it is advantageous for certain applications such as biodegradable implants. Since the 18th century, when Au, Ag and Pt elements were used for the fabrication of biomaterials [1], a large number of alloys have been developed so far. Some of the most employed metallic biomaterials for permanent implants are austenitic steels [2], Co-Cr-Mo [3], titanium and Ti-6Al-4V alloys [4] due to their biocompatibility and adequate mechanical behavior. To avoid post-extraction of these materials, intensive efforts have been made in recent years to develop the so-called "biodegradable implants". The materials of choice for biodegradable metallic implants are iron-based [5] and Mg-based alloys [6] owing to their relatively fast biodegradability. From the point of view of the mechanical performance, Mg alloys are preferred because their stiffness (i.e., Young's modulus) is closer to that of human bone [7].

Since "biodegradable implants" become reabsorbed by the human body after a certain period of time, they should be composed of biocompatible alloying elements. For this reason the potential cytotoxicity of the constituent elements of an implant material has to be seriously considered at an early stage of material development. For example, elements such as Ni, Al, Cr and V are not suitable to be in contact with human tissues [8]. Their substitution by non-toxic elements such as Zn and Ca has permitted the fabrication of biocompatible Mg-based alloys with potential use as biomaterials. However, the problem with some Mg alloys is their exceedingly high corrosion rates in physiological conditions, which makes their biodegrada-bility to be faster than the time required to heal the bone [9]. For this reason it is important to decrease their degradation rate, and to keep their mechanical integrity until the bone heals. Another drawback of magnesium and its alloys is that corrosion is accompanied by intense hydrogen evolution. This gas can be accumulated in pockets next to the implants or can form subcutaneous gas bubbles.

This book chapter deals with the fundamental aspects of corrosion of magnesium based alloys in bodily fluids and reviews the various techniques that can be used to tune their degradation rate. The time-dependent evolution of their mechanical properties during the biodegradation process is also outlined.

2. Basic aspects of corrosion

Corrosion is a surface phenomenon greatly influenced by different media-related factors (chemical, electrochemical and physical) in which the material is placed. The corrosion behavior of Mg in aqueous environments proceeds by an electrochemical reaction with water to yield magnesium hydroxide $Mg(OH)_2$ and hydrogen gas [10]:

$$\text{Anodic reaction: } Mg \rightarrow Mg^{2+} + 2e \tag{1}$$

$$\text{Cathodic reaction: } 2H_2O + 2e^- \rightarrow H_2(g) + 2OH^- \tag{2}$$

$$\text{Overall reaction: } + 2H_2O \rightarrow Mg(OH)_2 + H_2 \tag{3}$$

The hydroxide anions generated through the cathodic reaction cause an increase of the pH of the solution [11] (eq. (2)). The formation of a magnesium hydroxide $Mg(OH)_2$ layer onto the

Mg surface can further protect the metal from ongoing corrosion provided that the electrolyte pH and/or the presence of chloride anions or other species induce breakage of the passive film. According to the potential-pH Pourbaix diagram for magnesium in pure water at 25°C (Fig. 1), a passivation region exists for pH values above 10.4 [12] (alkaline environment) where the $Mg(OH)_2$ layer is stable. In neutral or acid environments (pH lower than 10.4) this layer is unstable. The diagram also shows that the immunity region of the diagram is below the region of water stability. However, bodily fluids are more aggressive than pure water. Body fluids are complex saline solutions containing ingredients such as proteins, blood serum, etc [13]. The most common fluids to carry out in-vitro tests and thereby to predict the degradation rate of magnesium and its alloys are Hank's balanced salt solution (HBSS), phosphate buffered solution (PBS) and simulated body fluid (SBF). All of them are acellular isotonic solutions (i.e., solutions with the same salt concentration as blood and cells) to make the sample, cell or tissue stable during an experiment.

HBSS [14] is mainly composed of chloride, sodium, potassium, magnesium and calcium ions. However, there are varieties of ingredients which can consist of glucose, potassium chloride (KCl), potassium dihydrogen phosphate (KH_2PO_4), sodium dihydrogen phosphate (NaH_2PO_4), and sodium chloride (NaCl). Additional ingredients can include hydrated magnesium sulfate ($MgSO_4\ 7H_2O$) and sodium bicarbonate ($NaHCO_3$).

PBS as its name implies [15] is a buffer solution consisting of a mixture of a weak acid and its conjugate base or a weak base and its conjugate acid. It aims to maintain a neutral pH in order not to destroy the cell or tissue sample and to maintain the osmolarity of the cells. The main ingredients are sodium phosphate and sodium chloride (NaCl) but in some recipes potassium phosphate and potassium chloride (KCl) are added.

SBF is a solution that has an inorganic ions concentration and pH almost equal to that of human extracellular fluid (i.e., the human blood plasma). The ions concentration in SBF is: Na^+ (142.0), K^+ (5.0), Mg^{2+} (1.5), Ca^{2+} (2.5), Cl^- (148.8), $HCO3^-$ (4.2) and $PO4^{2-}$ (1.0) mmol/dm^3 and it is buffered at pH 7.25 [16].

Chloride ions are able to dissolve the $Mg(OH)_2$ layer [17] yielding the soluble $MgCl_2$ salt [18], according to the following reaction:

$$Mg(OH)_2 + 2Cl^- \leftrightarrow MgCl_2 + 2OH^- \qquad (4)$$

Chloride ions are thus detrimental for the corrosion resistance of passive systems. Yet, other studies point out to opposite effects. For example, chloride ions were found to improve surface stability of Mg-Y-RE alloy in artificial plasma solution [19]. Other species can also degrade the protective passive characteristics of $Mg(OH)_2$ layer. Baril and Pébère found that the addition of increasing concentrations of $NaHCO_3$ to a deaerated Na_2SO_4 media leads to an accelerated corrosion of magnesium due to dissolution of MgO and $Mg(OH)_2$ films [20]. On the contrary, certain anionic species like HCO_3^- have beneficial effects and can be added to the electrolyte to increase the stability of the corrosion

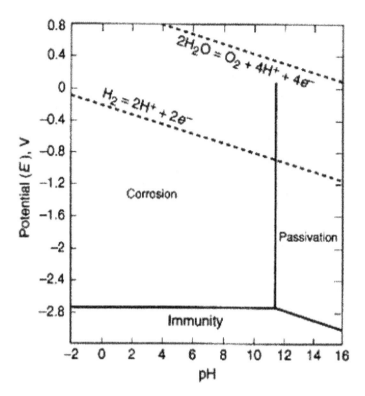

Figure 1. Potential-pH Pourbaix diagram for Mg in water at 25 °C.

products.The presence of dissolved O_2 appears not to play a major role in the corrosion of magnesium when immersed in saline solutions or fresh water [21].

3. Hydrogen evolution

One of the major drawbacks of Mg as biomaterial is the formation of H_2 gas when it is in contact with body tissues. The evolved H_2 bubbles from magnesium implants can be accumulated and form gas pockets that may lead to necrosis of the neighboring tissues and delay healing of the surgery region [22]. However, if the H_2 gas is generated slowly enough it can be transported away from the implant and can thus be tolerated by the body. According to Song [22] a hydrogen release rate in the human body of 0.01 ml/cm²/day can be tolerated. Dissolution of Mg and concomitant hydrogen evolution can be retarded by either purification of Mg or through appropriate alloying. Fig. 2 shows the average rate of hydrogen evolution (ml/cm²/day) for commercially pure Mg (CP-Mg) and different Mg alloys [22]. The highest release of hydrogen stands for CP-Mg, about 26 ml/cm²/day, and decreases with the addition of certain elements. For example, it decreases to 1.502 ml/cm²/day for ZE41 alloy (4 wt. % Zn, 1 wt. % RE), to 0.280 ml/cm²/day for Mg1.0Zn (1.0 wt. % Zn), to 0.068 ml/cm²/day for AZ91 (9 wt. % Al, 1 wt. % Zn) and to 0.012 ml/cm²/day for Mg2Zn0.2Mn (2 wt. % Zn, 0.2 wt. % Mn).

By measuring the hydrogen evolution rate the corrosion rate associated with magnesium is directly obtained since the release of one mol of H_2 implies the consumption of one mole of Mg according to eq. (3) [23]. The rate of H_2 gas evolution for Mg in Hank's solution at 37°C and different pH values was studied by Ng et al. [23] over a period of 7 days. They reported that the hydrogen evolution rate decreases with the increase of the solution pH. However, the volume of H_2 gas evolved over the time at a given pH (between 5.5 and 6.8) practically does not change. The same authors reported that the average H_2 evolution rate initially drops very fast from 153.3 to 1.079 ml/cm²/day when the pH rises from 5.5 to 7.4 but it slows down at pH 8.0 (0.534 ml/cm²/day) [23]. This was attributed to the accumulation of corrosion products that covered the sample surface, forming a progressively thicker layer with pH. Similarly, Zainal Abidin et al. [24] suggested that the formation of a partially protective film on Mg2Zn0.2Mn and ZE41 samples after long immersion times in Hank's solution was responsible for the decrease of the corrosion rate and concomitant H_2 evolution.

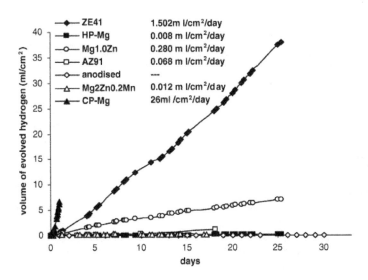

Figure 2. Hydrogen evolution in SBF and their average rates for various Mg-based alloys. Reprinted from Song G [22], page 3, with permission from Elsevier.

It is important to stress that magnesium shows an unusual electrochemical phenomenon known as "negative difference effect" (NDE) [25], which basically consists of an increase of the H_2 evolution rate at more positive potentials. For most metals, hydrogen evolution decreases with an increase of the applied potential or current density [26].

4. Corrosion of Mg and Mg alloys

When Mg and its alloys are used as biomaterials for implant applications they can be subjected to a combination of corrosion and stress (erosion, fatigue, etc). Since galvanic and pitting corrosion are the most common corrosion types of Mg and Mg alloys, this chapter primarily focuses on them:

4.1. Galvanic corrosion

Galvanic corrosion is an electrochemical process that occurs when two metals having different electrochemical potentials are in close contact with a common electrolyte. Of these two metals, the one that is more active in the galvanic series corrodes preferentially. Fig. 3 shows the galvanic series of different alloys listed in the order of the potential they exhibit in flowing seawater [27]. The black boxes of Fig. 3 correspond to the potentials in low-velocity or poorly aerated water. The reference potential is the Standard Calomel Electrode (SCE).

Although the composition of seawater differs slightly from that of saline body fluid and thus the corrosion potential is not expected to be exactly the same, Fig. 3 already gives a rough idea of the activity of different metals and alloys. The most positive (noble) material will be protected against corrosion at the expense of the material with more negative potential. Since the electrochemical potential of Mg and its alloys is located at the most negative side of this series (i.e., below -1.6 V), almost all the other metals in contact with it will be cathodically protected. Therefore, Mg will undergo galvanic corrosion; i.e., galvanic couples between the Mg metal or its alloy and the other metal will result in the dissolution of the former. The driving force for the galvanic corrosion depends on the difference between the potential (i.e., nobility) of both materials.

Regrettably, the corrosion of Mg alloys not only occurs when they are in close contact with other metals but also within the material itself. Mg alloys do not normally have a uniform microstructure, composition and crystalline orientation. This lack of uniformity is sufficient to promote the occurrence of galvanic couples [28]. The galvanic effect depends on a variety of factors; the crystal orientation of the matrix phase (i.e., the continuous phase of pure Mg into which the second phase/s is/are embedded), the alloying element concentrations in the matrix phase, the type and concentration of secondary phases along grain boundaries and the type and concentration of impurity particles in the matrix phase [28]. In the following, the main features having an influence of the corrosion rate of Mg are summarized:

- **Crystal orientation of the matrix phase:** polycrystalline pure Mg matrix immersed in neutral 0.01 M NaCl solution is more stable and corrosion resistant when grains possess a basal orientation [29]. This behavior can be explained considering that densely packed crystallographic planes (i.e., basal planes) normally have a higher atomic coordination and thus a lower dissolution tendency than non-compact planes [30]. For this reason, by controlling surface texture, one can improve the corrosion resistance of the material. For example, by rolling an AZ31 alloy it is possible to orient most of the crystallographic basal planes of the grains parallel to the rolling surface and thus decrease the corrosion rate of the rolled surface [31].

- **Alloying element concentrations:** the corrosion behavior of Mg phase can be tuned as a function of the concentration of elements in solid solution. Depending on the nature and distribution of these elements within the matrix phase, the occurrence of micro-galvanic cells can be either mitigated or favored. For example, an Al-containing Mg matrix phase becomes more passive as the Al content increases and consequently the corrosion rate decreases [32]. In as-cast Mg-Al alloys, the aluminum content in solid solution can vary from 1.5 wt. % at the grain center to about 12 wt. % at the grain boundary due to segregation during solidification [33]. Since Al has higher potential (Fig. 3) than Mg, corrosion mainly occurs at the interior of Mg grains. On the contrary, in Zr-containing Al-free Mg alloys, the central areas of the grains (which are enriched in Zr) do not corrode while the grain boundaries severely corrode.

- **Type and concentration of secondary phases along grain boundaries:** Mg intermetallic phases are typically nobler than the Mg matrix. As a consequence, they act as micro-galvanic cathodes and the dissolution of the Mg matrix is accelerated. Yet, in some cases the inter-metallic phases can stop the corrosion process. Hence, they actually play a dual role in the corrosion of Mg alloys [34]. Namely, the presence of a finely and continuously distributed secondary phase can stop the corrosion process while the presence of small amount of discontinuous secondary phase particles will accelerate it [25,35].

- **Type and concentration of impurity particles within the matrix phase:** the corrosion resistance of Mg alloys can be improved by limiting the concentration of critical impurities. However, not all the impurity elements have the same effect on the corrosion behavior. Some of them have little influence while others are very detrimental to the corrosion resistance. For example, Zn and Ca, which are frequently employed in the biomaterials field [6,36], have moderate accelerating effects on corrosion rates. Contrarily, Ni, Fe, Cu and Co are deleterious due to their low solid-solubility limits in Mg and their ability to act as cathodic sites [37]. The corrosion rate also depends on the impurity concentration. Each impurity has a tolerance limit. For impurity concentrations lower than the tolerance limit, there is no significant influence on the corrosion rate, whereas above this limit the corrosion rate sharply increases (Fig. 4) [37]. There is a rough relationship between the solubility of some elements in Mg alloys and their critical concentrations [38]. For example, the tolerance limit of Fe in Mg corresponds to the solubility of Fe in Mg [39].

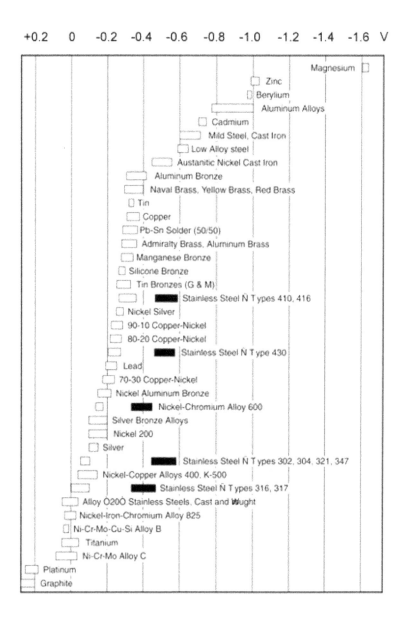

Figure 3. The galvanic series of metals, semi-metals and alloys of industrial interest showing their potentials (in volts) in flowing sea water, arranged from the most noble (bottom) to the most active (top) material. The values are referred to saturated calomel half-cell reference electrode. Adapted and reprinted from Amtec Consultants [27]

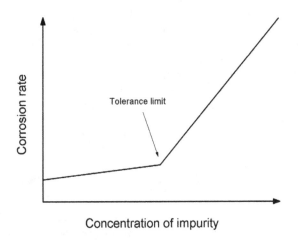

Figure 4. Schematic picture showing the dependence of the impurity concentration on the corrosion rate of Mg. The tolerance limit sets the threshold between the region for which an increase of the impurity concentration hardly affects the corrosion rate (left) and the region for which a further increase of the impurity concentration abruptly increases the corrosion rate.

• **Amorphous versus crystalline microstructure:** When a liquid is cooled below its liquidous temperature, it either crystallizes or, if crystallization is suppressed, it forms an amorphous solid. The microstructure and constituency of a material can be altered on purpose by means of rapid solidification processing at quenching rates of 10^5-10^6 K/s [40]. The development of novel Mg alloys with higher glass-forming ability has permitted to obtain amorphous materials with lower critical cooling rates. Since amorphous materials usually exhibit better corrosion and wear resistance than their crystalline counterparts, they can be potentially used for biomedical applications. Moreover, by controlling the crystallization events from the early stages of solidification, it is possible to tune the microstructure (i.e., nature and size of crystalline phases) and, in turn, optimize the corrosion performance of the material. This issue will be deeply tackled in section 5.1.2.

The micro-galvanic corrosion is also dependent on the solution in which the alloy is immersed. In a 3 % NaCl solution, the secondary phases present in the AZ91, ZE41 and Mg2Zn0.2Mn alloys can accelerate the corrosion rate. On the contrary, they do not play an important role when these alloys are immersed in Hank's solution [24]. The driving force for micro-galvanic corrosion between α-Mg and the secondary phases can be alleviated with the formation of a protective surface film on ZE41 and Mg2Zn0.2Mn during long immersion times in Hank's solution [24].

4.2. Pitting corrosion

It is a type of corrosion in which there is an intense localized attack on sample surface that leads to the formation of small holes in the metal. Mg alloys are prone to pitting corrosion when the passivation layer (which consists of $Mg(OH)_2$ [41] or a mixture of MgO and $Mg(OH)_2$ [42,43]) breaks down locally. When this occurs, the corrosion can be initiated at these local sites that act as small anodic surface areas. As aforementioned (see section: Basic Aspects of Corrosion), this protective coating may be damaged in physiological solutions because it is sensitive to both the chloride ion concentration and the solution pH. Physiological environments also contain phosphates, carbonates and sulfates that have different effects on the degradation behavior of Mg. Sulfate ions appear to stimulate the corrosion of Mg [44] while phosphate ions can delay pitting corrosion. Finally, carbonate ions can favor surface passivation and inhibit chloride-induced pitting corrosion due to the precipitation of stable magnesium carbonates into the pits. The presence of ions in different concentrations may explain why corrosion in Hank´s solution is more severe than in simulated blood plasma [45]. Nevertheless, immersion of AZ91, ZE41 and Mg2Zn0.2Mn alloys in Hank´s solution favors the formation of a more protective surface film than in 3% NaCl solution.

The large influence of the electrolyte composition on the corrosion behavior could explain why the same alloy exhibits different modes of corrosion depending on the environment. For example, a commercial AZ91 alloy immersed in 1M NaOH solution for 1 h and then re-immersed in 0.01M NaCl for 3h exhibited pitting [46] whereas the same alloy immersed in Minimum Essential Medium (MEM) at 37°C and 5 % CO_2 exhibited general corrosion mode [46]. Pitting can be also initiated by small surface defects such as scratches [47]. While galvanic corrosion is caused by local change of composition, pitting appears to be mainly influenced by the formation of a partially protective film. In fact, for the AZ91, ZE41 and Mg2Zn0.2Mn alloys, which are two-phase Mg alloys, their corrosion rates in Hank´s solution are similar to that of HP-Mg despite the tendency of the second phase to accelerate the corrosion rate [24]. The formation of a more protective and compact film on the surface of the Mg-Nd-Zn-Zr alloy than on AZ31 alloy is responsible for the slower corrosion rate on the former alloy in Hank´s solution at 37°C for 240h [48]. After immersion, deep pits were detected on the surface of the AZ31 alloy while that of the Mg-Nd-Zn-Zr alloy remained smooth.

The corrosion rate of Mg alloys is also influenced by the flowing conditions of the physiological environment (i.e., under dynamic physiological conditions the corrosion rate would slow down compared with steady conditions). This behavior is attributed to the fact that if the Hank´s solution is flowing, the absorption of Cl ions on the surface of the protective layer would be hindered. This phenomenon could explain the difference between in vitro and in vivo corrosion of degradable Mg alloys. For example, corrosion tests of AZ91D and LAE442 alloys in physiological solution indicate that both alloys corrode about four orders of magnitude slower in vivo than in vitro [49].

5. Tuning the biodegradation rate

The main limitation of Mg alloys for their use as implant materials is their exceedingly high corrosion rate in physiological conditions (i.e., pH = 7.4-7.6 and large chloride concentrations), which causes their biodegradability to be faster than the time required to heal the bone [50].

For this reason it is important to decrease the degradation rate of Mg alloys to remain implanted in the human body for at least 12 weeks [51]. Moreover, although the human body strives to keep a constant value of the pH, the presence of a fast corroding Mg implant can lead to local alkalinization that would unfavorably affect the pH close to the implant. Song suggested that a pH higher than 7.8 can have a poisoning effect [23]. These drawbacks associated with an exceedingly fast degradation rate suggest the need to control the biodegradation rate of Mg alloys.

So far, two kinds of methods have been used to slow down the corrosion rates of Mg alloys:

a. Surface coatings or surface modification:

In a broad sense, coatings can be divided in two classes: conversion coatings and deposited coatings. Conversion coatings consist of protective layers prepared using chemical (immersion in chemical baths to form calcium phosphate-containing layers, fluoride-containing layers, etc) or electrochemical processes (passivation, anodization, etc) [52]. Likewise, deposited coatings can be divided into metallic [53-56], organic [57] and inorganic [58,59]. The corrosion resistance of Mg and Mg alloys can be also improved through surface modification using various techniques such ion implantation [60], surface cladding and melting with laser or electron beam, [61], plasma surface modification [62], surface amorphization [6], etc.

b. By controlling the composition and the microstructure:

Although there are numerous review articles dealing with the growth of coatings [52] and surface modification procedures for Mg alloys [63], none of them deeply focuses, to the best of our knowledge, on the different means to tune the corrosion rate of these materials based on the control of their microstructure and composition. For this reason, the following section focuses on this subject.

5.1. Compositional and microstructural control:

5.1.1. Microstructural modification and thermal treatment

Microstructural control is an effective means to tune the strength and corrosion resistance of Mg alloys. More grain boundaries that act as corrosion barriers [64,65] are formed when the grain size is reduced. The microstructure can be refined using different severe plastic deformation (SPD) methods such as extrusion and equal channel angular extrusion (ECAP). Subsequent heat treatments further allow controlling the microstructure in order to tune the mechanical and corrosion performance. There are, in fact, multiple combinations of SPD and/ or heat treatments to optimize the microstructure. For example, an alloy can be first heat treated and then plastically deformed or an alloy can be simply heat treated from the as-cast condition (i.e., without undergoing plastic deformation).

An example of microstructural optimization through heat treatments is the effective control of the corrosion behavior of the as-cast (F) Mg3Nd0.2Zn (wt. %) (NZ) and Mg3Nd0.2Zn0.4Zr (wt. %) (NZK) alloys through solution heat treatment (T4) and solution heat treated and artificially aged (T6) in 5 % NaCl solution. The T4 treatment is carried out at 540°C for 6 h followed by water quench at 25°C. After this solution treatment the alloys are artificially aged in an oil bath at 200°C for 16h (T6) [66]. Immersion tests indicate that the highest corrosion rates stand for the as-cast samples: 1.353 mg/cm^2/day and 0.203 mg/cm^2/day for NZ and NKZ alloys, respectively. The heat treatments increase the corrosion resistance in the following order: F < T6 < T4. The lowest corrosion rates values are obtained at T4 conditions (0.266 mg/cm^2/day for NZ alloy and 0.11 mg/cm^2/day for NZK alloy) and only increase slightly at T6 conditions. The change in the corrosion rate after the heat treatments is ascribed to microstructural modifications. In the as-cast condition the microstructure of both alloys consist of α-Mg matrix and an eutectic Mg$_{12}$Nd compound inhomogeneously distributed within the matrix. Because of the discontinuous distribution, the Mg$_{12}$Nd acts as a microgalvanic cathode and, so, it accelerates the corrosion of the matrix. The authors reached this conclusion by comparing the role of Mg$_{12}$Nd phase on the corrosion behavior of NZ and NZK alloys with the role of β phase (i.e., Mg$_{17}$Al$_{12}$) on the corrosion of AZ alloys [27]. Song and Atrens proposed that when the β phase is discontinuously distributed within the material, the corrosion rate of AZ alloys increases (see section 5.1.3) and thus the same behavior is expected for the Mg$_{12}$Nd phase. However, when the NZ and NZK alloys are subjected to T4 or T6 treatments, the Mg$_{12}$Nd compound dissolves into the matrix and microgalvanic couples are no longer present. The slightly higher corrosion rates detected at T6 to T4 is attributed to the precipitation of very small Nd-rich precipitates. Consistently, the corrosion morphologies reveal that the localized attack zones are more severe in the as-cast than in the T4 condition. Also, in the T6 condition the attack is slightly more severe than in T4 condition.

For a ZE41 alloy (4 wt. % Zn, 1 wt. % RE) the corrosion behavior improves after heat treating for 5 days at 500°C [67]. This improvement is again related to microstructural changes that occur during heat treatment. The Mg$_7$Zn$_3$RE phase present in the material before heating partly redissolves, which explains the increasing concentration of Zn and RE in the matrix. Similarly, the corrosion resistance of the as-cast Mg10Gd3Y0.4Zr (wt. %) alloy increases with solution treatments due to the dissolution into the α-Mg matrix of the (Gd+Y)-rich eutectic present in the as-cast condition [68]. The improvement of the corrosion resistance greatly depends on the thermal treatment. Namely, it is highest for a T4 solution treatment (500°C for 6 h) than for any of the T6 solution treatments (oil bath at 250°C for 0.5, 16, 193 and 500 h). The reason lies in that an increasing ageing time increases the volume fraction of secondary phase that act as cathodes and thereby ultimately increases the corrosion rate.

The microstructure of alloys can be optimized if the temperature is properly controlled during the dynamic recrystallization in an extrusion process. The corrosion behavior in SBF at 37°C of Mg3Nd0.2Zn0.4Zr (wt. %) NZK alloy initially solution-treated at T4 conditions (at 540°C for 10 h and then water quenched to room temperature) is effectively modified by controlling the extrusion temperature (250°C, 350°C and 450°C) [69]. Both immersion and electrochemical tests indicate that the corrosion rate in the extruded condition at 250°C, 350°C and 450°C is much slower than in the T4 state. Moreover, the corrosion resistance increases with the decrease of the extrusion temperature and so does the grain size.

Deformation processing can also have an effect on the redistribution of solutes within the microstructure and ultimately affect the corrosion behavior. When a ZK60 (6 wt.% Zn, 0.5 wt.

% Zr) alloy is processed by an integrated extrusion combined with ECAP, it is observed that Zn-Zr and Mg-Zn intermetallics become fractured and redistributed within the microstructure. Electrochemical and immersion tests in NaCl electrolytes indicate that grain refinement and redistribution of Zr and Zn solutes improve the corrosion resistance [70].

The corrosion behavior also depends on microstructural effects such as twins, dislocations, etc., caused by deformation processing. For example, the corrosion resistance in 3.5 % NaCl of AZ31B magnesium alloy has been studied in the initially hard rolled condition and after heat treating at 200, 300, 400 and 500°C for 3 h in an inert atmosphere of argon and subsequent quenching in water to room temperature [71].

The initial average grain size of 35 µm in the as-received condition increases with the heat treating temperature to 50 µm at 200°C (HT 200), 65 µm at 300°C (HT 300), 90 µm at 400°C (HT 400) and to 250 µm at 500°C (HT 500). In the HT300 conditions, the microstructure is untwined because a high density of twins are eliminated and so the intra-granular corrosion is the least. However, in the as-received and HT200 conditions the deformation twins and thus the dislocation density is higher. This can explain the more serious corrosion of the HT200 microstructure compared with the HT300 microstructure despite the fact that the HT200 microstructure is finer and thus the physical corrosion barrier is larger. In other words, twins accelerate the corrosion. From the physical metallurgy viewpoint, in the as rolled conditions (i.e., after plastic deformation), the amount of twins is the largest but they are progressively annihilated as temperature increases. For this reason potentiodynamic polarization curves show that the corrosion rate increases as the microstructure becomes more twinned.

Not only the presence of twins but also the distribution and density of dislocations are correlated with the corrosion behavior. The AZ31 alloy plastically deformed by ECAP at 350°C with a pressing speed of 350 mm/min exhibit twins and a higher density of dislocations than after being extruded at 350°C with an extrusion ratio of 10.24 (in this case twins were not observed) [65]. From corrosion studies in 3.5 % NaCl saturated with $Mg(OH)_2$ at pH 10.5, the authors concluded that the corrosion rate of AZ31 alloy decreases after extrusion but it increases after ECAP, suggesting that the twins and/or presence of higher density of dislocations decisively affect the corrosion rate.

The corrosion behavior of a AZ31 Mg alloy with different grain sizes immersed in two different solutions, NaCl and phosphate-buffer solution (PBS) has been studied by other researchers. The microstructure is refined by ECAP with a first pass of 250°C and successively heat treated to 300°C and rolled [72]. The best corrosion behavior is attained by the alloy having finest grains after long-term immersion in PBS [72]. This behavior is related to the formation of a mixed compact protective layer of P-containing compounds together with magnesium hydroxide that promotes protection against the chloride ions. The superior corrosion behavior of the fine-grained AZ31 alloy over the coarser one is attributed to the formation of a layer of corrosion products that provides better protection against the diffusion of aggressive ions to the surface of the material [72].

Although these results suggest that the corrosion performance can be tuned by controlling the microstructure, other factors such as the chemical composition plays a more important role.

For example, Liao et al. [73] observed that the fine grained AZ31B alloy exhibits a lower corrosion resistance than the AM60 alloy with coarser grains.

5.1.2. Amorphous and partly amorphous alloys

As aforementioned, the grain size can be tuned by controlling the cooling rate. For certain compositions such as AZ91 alloy [74] rapid cooling is an effective technique to obtain fine grain sizes. For other Mg-based compositions a sufficiently fast cooling rate can lead to the formation of glassy materials. Moreover, rapid cooling allows to extend the solubility of alloying elements in Mg alloys and to form a homogeneous single-phase structure (i.e., metallic glass) with a very different corrosion behavior than that of crystalline Mg alloys [75]. Typically, amorphous materials possess stronger corrosion and chemical resistance than their crystalline counterparts due to the absence of grain boundaries, segregated phases, secondary particles and also due to chemical homogeneity [76]. For this reason different Mg-based glassy materials have been studied over the years. For example, glassy $Mg_{60+x}Zn_{35-x}Ca_5$ (0<x<7 at. %) ribbons of 50 μm in thickness can be successfully obtained by melt spinning [36]. Immersion tests of these ribbons in SBF lead to the formation of corrosion layers that are different from those found in Zn-poor and Zn-rich alloys. For the Zn-rich alloys(above 28 at. % Zn), the Zn-rich oxygen-containing passivating layer that is formed on the surface of the ribbon is responsible for the more noble behavior of these alloys as compared to the Zn-poor alloys [36]. Moreover, a high Zn content appears to reduce hydrogen evolution. In fact, due to the extended solubility of Zn in the amorphous structure of the Mg-Zn-Ca system, the $Mg_{60}Zn_{35}Ca_5$ glass only exhibits marginal hydrogen evolution during in vitro and in vivo degradation [36].

Through the addition of different alloying elements to the Mg-Zn-Ca alloys family, the corrosion behavior can be tuned as well. Small Pd additions are enough to decrease the glass forming ability of glassy $Mg_{72}Zn_{23}Ca_5$ alloys and to shift the corrosion potential towards more positive values [6]. Cytotoxic tests do not show the presence of death cells, which confirm that these alloys might have potential use as implants [77]. Cytocompatibility tests also show that metallic glass $Mg_{66}Zn_{30}Ca_4$ and $Mg_{70}Zn_{25}Ca_5$ samples have higher cell viability and exhibit more positive corrosion potential than that of as-rolled crystalline pure Mg [78].

It is well known that glassy materials can be used as precursors of crystalline phases by controlling the crystallization temperature and/or time. Since the corrosion behavior depends on the structure (i.e., amorphous vs. crystalline) of the material, the extent of crystallization can be controlled to tune the corrosion rate. For example, glassy $Mg_{67}Zn_{28}Ca_5$ ribbons exhibit an increase of the corrosion resistance in simulated body fluid with the increase of annealing temperature up to a maximum and then the resistance decreases rapidly for higher temperatures. The best corrosion resistance of these ribbons is attained at 160°C, when the microstructure is constituted by a metastable crystalline $Mg_{102.08}Zn_{39.6}$ phase embedded in an amorphous matrix [76]. This behavior was explained considering that the electrochemical activity of this phase is similar to that of its amorphous matrix. However, the newly formed phases at 225°C are more active and worsen the corrosion resistance of the alloy [76].

To determine the effect that alloying elements have on the corrosion resistance of rapidly solidified magnesium alloys, different binary Mg-based glassy alloys were studied by using electrochemical techniques in pH 9.2 sodium borate [79]. These studies concluded that the corrosion rate of magnesium is decreased for larger contents of aluminium. Similarly, low concentrations of zinc and lithium decrease the corrosion rate below that of pure magnesium [79]. These results indicate that composition has an important influence on the corrosion rate of glassy Mg alloys, as it has also been observed in crystalline alloys.

5.1.3. Influence of alloying elements on corrosion performance

As was explained on section 4.1, the corrosion rate of magnesium alloys depends on the nature and concentration of impurities. The corrosion resistance can be improved either by purifying Mg or through appropriate additions of alloying elements. Mg alloys are basically classified [34] in two groups: 1) those that contain Al as primary alloying element and 2) those that do not contain Al and have small amounts of Zr to refine the microstructure. Al is generally considered as a beneficial element to improve the corrosion resistance [80]. For small contents, Al remains in solid solution, but above certain concentration β-$Mg_{17}Al_{12}$ secondary particles precipitate.

The β-phase is very stable in NaCl solutions and it is inert to corrosion due to the formation of a passive thin film on its surface. However, β-phase is also an effective cathode, which can explain the dual role of these precipitates in the corrosion of AZ alloys according to Song and Atrens [25]. A fine and continuous distribution of β-phase is recommended to increase the corrosion resistance. For example, Lunder et al. reported that Al additions higher than 8 wt. % increase the corrosion resistance of Mg-Al alloys [81]. An improvement of the corrosion resistance with the Al content is also found in AZ91, AZ61 and AZ31 alloys in 5 % NaCl [82]. However, Song et al. reported an increase of the corrosion rate in NaCl in the following order AZ501 < AZ21 < AZ91 [83].

For the second family of alloys, small additions of Zr refine the microstructure of Mg and improve the corrosion resistance [84]. Since Zr can easily combine with impurities, especially Fe and Ni, it can form insoluble precipitates that settle out during melting. This purification effect of Zr enhances the corrosion resistance of Mg [85]. Depending on the composition, a minimum concentration of Zr is required to observe such effect. For example, for Zr contents from 0 to 0.42 wt. % the corrosion resistance of Mg10Gd3Y (10 wt. % Gd, 3 wt. %) deteriorates, whereas for higher Zr contents, ranging from 0.42 to 0.93 %, the corrosion resistance improves. The distinct behavior is attributed to the differences in size and distribution of the Zr-rich particles [86]. An exceedingly larger amount of Zr can lead to precipitation of Zr within the matrix, which becomes detrimental for the corrosion performance of the alloy [34].

The addition of 1 wt. % of Al, Ag, In, Mn, Sn, Zn and Zr elements decrease the volume of evolved hydrogen gas, and thus decrease the corrosion rate, of Mg when immersed either in SBF or in Hank's solution [87]. On the contrary, the addition of 1 wt. % Y or Si have a deleterious effect on the corrosion performance.

Ca is an essential element to the body since it is a major component of human bones. For this reason, it has been used over the years to fabricate biocompatible Mg-based alloys. The concentration of Ca has, though, to be carefully controlled to avoid the precipitation of Mg_2Ca particles (that takes place for Ca contents ranging from 0.8 to 5 wt. % [88] or from 1 to 3 wt. % [89] depending on the system under study). These Mg_2Ca particles form micro-galvanic cells within the Mg matrix and accelerate preferentially the dissolution of the latter, worsening the corrosion resistance of the binary Mg-xCa alloy. For 1.5 wt. % Ca, a protective oxide layer of MgO and CaO is formed after heating to 500°C for 1h [90]. The influence of Ca on the corrosion behavior not only depends on its amount but also on the composition of the Mg-based alloy to which it is added. For example, the addition of 13 wt. % Ca increases the corrosion rate of AZ91D alloy (37 wt. % Al, 0.5 wt. % Zn) [91]. An improvement of the biocorrosion resistance is also detected when 0.2-0.4 wt. % Ca is added to a Mg-Si alloy since it refines the grain size and modifies the morphology of Mg_2Si phase [92]. The same holds when 1.6 wt. % Zn is added to Mg-Si alloy due to the modifications on the Mg_2Si phase morphology derived from the addition; namely from a coarse eutectic structure to a small dot or short bar shape [92]. Zn is an essential element to the human body and capable of decreasing the corrosion rate of pure Mg in small amounts. For example, the corrosion rate (measured in terms of volume of evolved hydrogen) of CP-Mg decreases from 26 ml/cm^2/day to 0.280 ml/cm^2/day with the addition of 1 wt.% Zn [22]. The addition of 6 wt. % zinc shifts the corrosion potential toward more cathodic values and decreases the in-vitro degradation rate of high purity Mg in SBF [93]. However, concentrations above the equilibrium solid solubility of Zn in Mg (i.e., 6.2 wt. %) [94] can lead to an increase of the corrosion rate in 3 % NaCl due to the formation of β-Mg_7Zn_3 phase in the magnesium matrix [95]. The introduction of Mn can help to decrease the corrosion rate of Mg-Zn alloys. Ahmed et al. [96] reported that adding Mn to a Mg-based alloy containing 4 to 8 wt. %. Zn decreases the dissolution rate of Mg. The corrosion rate of Mg2Zn0.2Mn (2 wt. % Zn, 0.2 wt. % Mn) in Hank's solution is also smaller than that of Mg1Zn (1 wt. % Zn) [22].

Other atypical alloying elements such as Y, Ce, Ti and Sc were reported to improve the corrosion performance when alloyed with Mg at a level below the solubility limit [97].

6. Biodegradation and mechanical integrity

The use of Mg alloys as weight-bearing implants requires that the material should have sufficient strength not only at the moment of being implanted but also when the alloy degrades over the time while remaining in contact with body fluids. It is important that implants keep their strength at least until the bone heals. For this reason different studies have been carried out to evaluate the mass loss and evolution of the strength over the implantation or immersion time [77].

According to Pietak et al. [98] the best technique to assess the degradation of Mg alloys is measuring the mechanical integrity as a function of the incubation time. Nevertheless, the measurement of the mass change [46] has been more frequently used. However, this procedure has several shortcomings due to the association of non-soluble degradation products that precipitate on the sample and obscure the mass loss [98].

The mechanical integrity can be evaluated using various mechanical tests (three-point bending, tensile tests, nanoindentation, etc). These tests can be performed under physiological conditions or in air.

From nanoindentation tests, Pellicer et al. [77] studied the evolution of the Young's modulus (E_r), hardness (H), and H/E_r ratio (which is an indirect measure of the wear resistance) of amorphous $Mg_{72}Zn_{23}Ca_5$ after different immersion times in HBSS as shown in Fig. 5. The same study was carried out on crystalline $Mg_{70}Zn_{23}Ca_5Pd_2$ alloy and the results were systematically compared. While the stiffness of both compositions decreases with the immersion time, hardness exhibits a more complex dependence, especially for the $Mg_{72}Zn_{23}Ca_5$ alloy. Namely, an increase is observed after short-term immersion, which was mainly attributed to the fast dissolution of Mg and the concomitant enrichment in Zn (Zn is mechanically harder, so solution hardening takes place). For longer immersion times, the dissolution progresses and the alloy not only undergoes surface chemical change but the surface is also physically modified with to the formation of flaws such as pores and corrugations, which cause a decrease of hardness [77]. The values of H/E_r increase from 0.053 for the as-cast $Mg_{72}Zn_{23}Ca_5$ alloy to 0.1 for $Mg_{72}Zn_{23}Ca_5$ after 2h immersion in HBSS at 37°C, thus indicating that the effect of porosity on the Young's modulus for short-immersion times is more noticeable than on hardness. These results are consistent with those observed in many other metallic alloys [99].

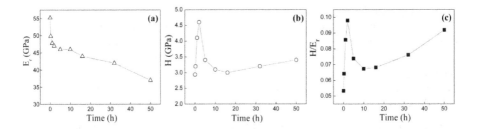

Figure 5. Dependence of (a) reduced Young's modulus, E_r, (b) hardness, H and (c) H/E_r ratio on the immersion time in HBBS at 37°C for $Mg_{72}Zn_{23}Ca_5$ alloy. Adapted and reprinted from Pellicer et al. [77], page 8, with permission from John Wiley&Sons.

To evaluate the mechanical integrity of AZ91Ca alloy (i.e., a calcium-containing magnesium alloy) in SBF at 36.5°C, slow strain rate tensile tests at 1.2×10^{-7} s^{-1} were carried out [91]. The AZ91Ca alloy shows lower elongation (3.5 ± 0.2 %) and lower ultimate tensile strength (106 ± 10 MPa) in the SBF than in air (4.6 ± 0.3 % and 126 ± 5 MPa, respectively). The decrease of the mechanical performance is, however, small and thus the alloy is not highly susceptible to corrosion in SBF. From electrochemical experiments it is observed that the AZ91Ca alloy exhibits improved corrosion resistance compared to the AZ91 alloy, which can be attributed to the formation of calcium phosphate on the surface of the AZ91Ca alloy. This surface film has higher stability than the film formed on AZ91 alloy for this reason not only the general corrosion resistance but also the pitting corrosion resistance improve.

Krause et al. [100] compared the evolution of the mechanical behavior of Mg0.8Ca (8 wt.% Ca), LAE442 (4 wt. % Li, 4 wt. % Al, 2 wt. % RE) and WE43 (4 wt. % Y, 3 wt. % RE) alloys implanted in rabbits for 3 and 6 months by three-point bending tests. All the samples exhibit biodegradation as can be deduced from the loss in volume with implantation period. The MgCa0.8 alloy degrades slowly during the first 3 months but its corrosion rate accelerates during the following 3 months. The LA442 alloy exhibits slower degradation rate than the Mg0.8Ca and WE43 alloys. The difference of degradation rate is responsible for the distinct mechanical performance of the alloys over the time. Thee-point bending test results indicate the following trend in the initial strength: LAE442 (255.67±5.69 N) > WE43 (238.05 ±21.68 N) > Mg0.8Ca (178.76±25.15 N). However, after 3 months the strength trend changes so that it decreases in the following order: WE43 (185.59±15.64 N) > LAE442 (153.21±18.45 N) > Mg0.8Ca (115.42±9.66 N). After 6 months the strength follows this sequence: LAE442 (134.68±14.68 N) > WE43 (122.23±23.65 N) > Mg0.8Ca (52.90±5.96 N). From the results of the maximal applied force it can be deduced that the LAE442 alloy degrades faster during the first 3 months and slower between 3 and 6 months. The degradation rate of Mg0.8Ca and WE43 alloys is different; it decreases in a linear manner over the time [100]. The ductility of the alloys was also assessed from three-point bending tests by measuring the bending displacement but concluding results could not be obtained due to high scattering. The Mg0.8Ca alloy exhibits the highest loss and the LAE442 the lowest loss in volume after 6 months.

The evolution of the bending strength of Mg6Zn (6 wt. % Zn) alloy with the immersion time in physiological saline solution (0.9 % NaCl) [88] is similar to that of implanted LAE442 alloy [100]. For short immersion times (i.e., 3 days) the degradation rate of Mg6Zn is very fast (0.20±0.05 mm/year) and exhibits a large weight loss but it becomes slower (0.07±0.02 mm/year) for longer immersion times (i.e., 30 days). The bending strength of the alloy decreases rapidly with an initially small weight loss but then decelerates as the percentage weight loss increases. This behavior was attributed to the formation of surface defects such as corrosion holes during degradation.

Plates of ZEK100 (1 wt. % Zn, 0.1 wt. % Zr, 0.1 wt. % RE (rare earth)) were mechanically tested in vitro after 14 (2 weeks), 28 (4 weeks) and 42 days (6 weeks) of immersion with a constant laminar flow rate in HBSS via four-point bending tests [101]. The bending strength decreases from immersion week 2 to week 4 but increases again after 6 weeks. The lowering of the bending strength is attributed to dissolution of the plate whereas the increase at longer times can be explained by precipitation of calcium phosphates from the solution on the surface of the plate. This behavior was supposed to be caused by a decrease of the implant volume during the first 4 weeks and an increase for longer times up to 8 weeks.

The mechanical behavior of ZEK100 alloy was also tested via three-point bending after being implanted in rabbit tibiae for 3 and 6 months [102]. The corrosion rate increases from 0.067 mm/year after 3 months to 0.154 mm/year after 6 months. The volume of the implant tends to reduce with the increase of the implantation time. This can explain why the initial maximum force of 241 N (the maximum force at breakage) decreases to 153 and 100 N after 3 and 6 months, respectively.

Figure 6 shows a comparative summary of the mechanical properties (compressive yield stress, $\sigma_{y,C}$, and Young's modulus, E) of various families of materials that can be used as bioabsorbable implants, such as metallic alloys, biodegradable polymers and ceramics. From the correlation $H \approx C \, \sigma_{y,C}$ (where C is the so-called constraint factor and is normally close to 3 for crystalline metallic alloys and slightly higher for metallic glasses [103]) the mechanical hardness is observed to be directly proportional to $\sigma_{y,C}$.

The yield stress of Mg-Zn-Ca bulk metallic glasses is relatively large (Figure 6), thus indicating that they are one of the hardest biodegradable materials reported in the literature. Moreover, the values of Young's modulus of glassy Mg-Zn-Ca are closer to that of cortical bone (E_{bone} = 3– 20 GPa) than most crystalline Mg-based alloys (they are also mechanically softer) and synthetic hydroxyapatites. Considering that the $Mg_{70}Zn_{23}Ca_5Pd_2$ alloy is fully crystalline, its hardness is also generally larger than that of most Mg-Zn-Ca crystalline alloys. Compared with most Fe-based biodegradable alloys, Mg-based bulk metallic glasses are generally harder. Moreover, Fe-based alloys are typically ferromagnetic at room temperature (except the antiferromagnetic FeMn), which precludes their use in nuclear resonance imaging techniques for diagnostics purposes. The Young's modulus of $Mg_{72}Zn_{23}Ca_5$ becomes closer to that of Ca-based or Sr-based biodegradable metallic glasses after long-term immersion in HBSS as well as to that of polymeric materials reinforced with glassy fibers. At the same time, the hardness of $Mg_{72}Zn_{23}Ca_5$ alloy is higher than that of all these materials.

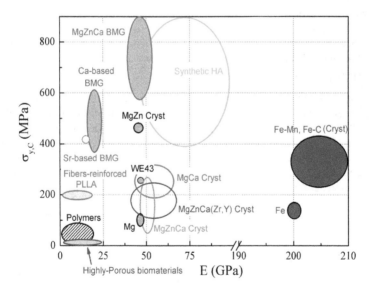

Figure 6. Comparison of the mechanical properties (compressive yield stress, $\sigma_{y,C}$, versus Young's modulus, E) in linear scale for different families of biodegradable implant materials, including metallic alloys (Mg-based, Ca-based, Sr-based, or Fe-based), ceramics (e.g., synthetic hydroxyapatites) and polymers. Adapted and reprinted from Pellicer et al. [77], page 13, with permission from John Wiley&Sons.

7. Summary and conclusion

Magnesium and its alloys are suitable materials for biomedical applications due to their low weight, high specific strength, stiffness close to bone and good biocompatibility. Specifically, because magnesium exhibits a fast biodegradability, it has attracted an increasing interest over the last years for its potential use as "biodegradable implants". However, the main limitation is that Mg degrades too fast and that the corrosion process is accompanied by hydrogen evolution. In these conditions, magnesium implants lose their mechanical integrity before the bone heals and hydrogen gas accumulates inside the body. To overcome these limitations different methods have been pursued to decrease the corrosion rate of magnesium to acceptable levels, including the growth of coatings (conversion and deposited coatings), surface modification treatments (ion implantation, plasma surface modification, etc) or via the control of the composition and microstructure of Mg alloys themselves.

For tuning efficiently the composition and microstructure it is first necessary to understand two of the most common types of corrosion that Mg and Mg alloys exhibit: galvanic and pitting corrosion. Galvanic corrosion develops because magnesium almost always behaves anodically in contact with other metals. Galvanic couples are usually encountered when the concentration of the alloying element surpasses their corresponding maximum solid solubility in magnesium. The alloying element then segregates during solidification or annealing and an inhomogeneous microstructure is formed. The extent of the galvanic effect depends on a number of factors such as the crystal orientation of the magnesium matrix, the type of secondary phases and impurity particles, the solution in which the alloy is immersed and the grain size. The concentration and distribution of secondary phases is also important for the corrosion behavior. A fine and continuous distribution of secondary phases typically improves the corrosion performance.

Mg alloys are susceptible to form a passivation layer of $Mg(OH)_2$ or a mixture of $Mg(OH)_2$ and MgO in aqueous solutions. Due to the presence of chloride ions in physiological fluids, the protective coating may be destroyed and localized attack (i.e., pitting corrosion) initiates. Physiological environments also contain carbonates, phosphates, sulfates and other ingredients that have different effects on the corrosion behavior of magnesium. Before magnesium alloys can be used as real implants it is necessary to evaluate the biodegradability and mechanical performance over the immersion (in-vitro) or implantation (in vivo) time.

Acknowledgements

This work has been partially financed by the 2009-SGR-1292 and MAT2011-27380-C02-01 research projects. S. G. acknowledges the Juan de la Cierva Fellowship from the Spanish Ministry of Science and Innovation. M.D.B. was partially supported by an ICREA Academia award.

Author details

S. González[1], E. Pellicer[1], S. Suriñach[1], M.D. Baró[1] and J. Sort[2]

*Address all correspondence to: Sergio.Gonzalez@uab.es

1 Departament de Física, Facultat de Ciències, Universitat Autònoma de Barcelona, Barcelona, Spain

2 Institució Catalana de Recerca i Estudis Avançats (ICREA) and Departament de Física, Facultat de Ciències, Universitat Autònoma de Barcelona, Barcelona, Spain

References

[1] Bhat SV. Biomaterials. Kluwer Adademic Publishers. Boston, MIT, USA; 2002. p265.

[2] Sumita M, Hanawa T, Teoh SH. Development of nitrogen-containing nickel-free austenitic stainless steels for metallic biomaterials-review. Matererials Science and Engineering C 2004; 24: 753-760.

[3] Puleo DA. Biochemical surface modification of Co-Cr-Mo. Biomaterials 1996; 17: 217-222.

[4] Geetha M, Singh AK, Asokamani R, Gogia AK. Ti-based biomaterials, the ultimate choice for orthopaedic implants. A review. Progress in Materials Science 2009; 54: 397-425.

[5] Moravej M, Mantovani D. Biodegradable metals for cardiovascular stent application: interests and new opportunities. International Journal of Molecular Sciences 2011; 12: 4250-4270.

[6] González S, Pellicer E, Fornell J, Blanquer A, Barrios L, Ibañez E, Solsona P, Suriñach S, Baró MD, Nogués C, Sort J. Improved mechanical perfomance and delayed corrosion phenomena in biodegradable Mg-Zn-Ca alloys through Pd-alloying. Journal of the Mechanical Behavior of Biomedical Materials 2012; 6: 53-62.

[7] Hanbook of Biomaterials Properties. In: Black J, Hasting GW (ed.). Chapman and Hall. London; 1998.

[8] Zhou Z, Liu X, Liu Q, Liu L. Evaluation of the potential cytotoxicity of metals associated with implanted biomaterials (I). Preparative Biochemistry and Biotechnology 2009; 39: 81-91.

[9] Li XN, Gu ZJ, Lou SQ, Zheng YF. The development of binary Mg–Ca alloys for use as biodegradable materials within bone. Biomaterials 2008; 29: 1329–1344.

[10] Godard HP, Jepson WB, Bothwell MR, Kane RL. The Corrosion of Light Metals. In: John Wiley&Sons (ed.). New York; 1967.

[11] Wang L, Shinohara T, Zhang B-P. Influence of deaerated condition on the corrosion behavior of AZ31 magnesium alloy in dilute NaCl solutions. Matererials Transactions 2009; 50: 2563-2569.

[12] Pourbaix M. Atlas of Electrochemical Equilibrium in Aqueous solutions. In: 2nd Ed. NACE. Houston; 1974.

[13] ASM Handbook. Volume 13A Corrosion: Fundamentals, Testing and Protection. In: Cramer SD, Covino BS, Jr. ASM International; 2003.

[14] Lonza Walkersville Inc. http://vgn.uvm.edu/outreach/documents/Hanksbufferedsalinesolution.pdf (accessed 27 November 2012).

[15] Medicago AB. Phosphate buffered saline specification sheet; 2010.

[16] Kokubo T, Kushtani T, Sakka S, Kitsugi T, Yamamuro T. Solutions able to reproduce *in vivo* surface-structure changes in bioactive glass-ceramic A-W3. Journal of Biomedical Materials Research 1990; 24: 721-734.

[17] Tunold R, Holtan H, Berge MBH, Lasson A, Steen-Hansen R. The corrosion of magnesium in aqueous solution containing chloride ions. Corrosion Science 1977; 17: 353-365.

[18] Hara N, Kobayashi Y, Kagaya D, Akao N. Formation and breakdown of surface films on magnesium and its alloys in aqueous solution. Corrosion Science 2007; 49: 166-175.

[19] Quach NC, Uggowitzer PJ, Schmutz P. Corrosion behavior of an Mg-Y-RE alloy used in biomedical applications studied by electrochemical techniques. Chemie 2008; 11: 1043-1054.

[20] Baril G, Pébère N. The corrosion of pure magnesium in aerated and deaerated sodium sulphate solutions. Corrosion Science 2001; 43: 471-484.

[21] Handbook of Corrosion data. In: Craig BD and Anderson D (ed.). Materials data series. Materials Park, OH: ASM International; 1995.

[22] Song GL. Control of biodegradation of biocompatible magnesium alloys. Corrosion Science 2007; 49: 1696-1701.

[23] Ng WF, Chiu KY, Cheng FT. Effect of pH on the in vitro corrosion rate of magnesium degradable implant material. Materials Science and Engineering C 2010; 30: 898-903.

[24] Zainal Abidin NI, Martin D, Atrens A. Corrosion of high purity Mg, AZ91, ZE41 and Mg2Zn0.2Mn in Hank's solution at room temperature. Corrosion Science 2011; 53: 862-872.

[25] Song GL, Atrens A. Corrosion mechanisms of magnesium alloys. Adv. Eng. Mater. 1999; 1: 11-33.

[26] Song GL, Atrens A, StJohn DH, Wu X, Nairn J. The anodic dissolution of magnesium in chloride and sulphate solutions. Corrosion Science 1997; 39: 1981-2004.

[27] Amtec Consultants. Experts in Coatings & Corrosion; 2011.

[28] Song GL. Corrosion characteristics of Mg alloys: NACE DOD Conference, July 2011, Palm Springs, CA.

[29] Song GL, Xu Z. Crystal orientation and electrochemical corrosion of polycrystalline Mg. Corrosion Science 2012; 63: 100-112.

[30] Suárez MF, Compton RG. Dissolution of magnesium oxide in aqueous acid: an atomic force microscopy study, Journal of Physical Chemistry B 1998; 102: 7156–7162.

[31] Song GL, Mishra R, Xu Z. Crystallographic orientation and electrochemical activity of AZ31 Mg alloy. Electrochemistry Communications 2010; 12: 1009-1012.

[32] Song GL, Bowles A, StJohn DH. Corrosion resistance of aged die cast magnesium alloy AZ91D. Materials Science and Engineering A 2004; 366: 74-86.

[33] Dargusch MS, Dunlop GL, Pettersen K. Mg alloys and their applications. Wolfsburg, Germany: Werkstoff-Information GmbH; 1998 p277-282.

[34] Song GL. Recent progress in corrosion and protection of magnesium alloys. Advanced Engineering Materials 2005; 7: 563-586.

[35] Liu C, Xin Y, Tang G, Chu PK. Influence of heat treatment on degradation behavior of bio-degradable die-cast AZ63 magnesium alloy in simulated body fluid. Materials Science and Engineering A 2007; 456: 350-357.

[36] Zberg B, Uggowitzer PJ, Löffler JF. MgZnCa glasses without clinically observable hydrogen evolution for biodegradable implants. Nature Materials 2009; 8: 887-891.

[37] ASM Specialty Handbook - Magnesium and Magnesium Alloys. In: Avedesian MM and Baker H (ed.). ASM international. Materials Park, OH; 1999.

[38] Roberts CS. Chapter Mg alloy systems. In: Mg and its alloys, John Wiley & Sons; 1960. p42-80.

[39] Liu M, Uggowitzer PJ, Nagasekhar AV, Schmutz P, Easton M, Song G-L, Atrens A. Calculated phase diagrams and the corrosion of die cast Mg-Al alloys. Corrosion Science 2009; 51: 602-619.

[40] Wang WH, Dong C, Shek CH. Bulk metallic glasses. Materials Science and Engineering R 2004; 44: 45-89.

[41] Badawy WA, Hilal NH, El-Rabiee M, Nady H. Electrochemical behavior of Mg and some Mg alloys in aqueous solutions of different pH. Electrochimica Acta 2010; 55: 1880-1887.

[42] Phillips RC, Kish JR. On the self-passivation tendency of Mg-Al-Zn (AZ) alloys in aqueous solutions. ECS Transactions 2012; 41: 167-176.

[43] Yao HB, Li Y, Wee ATS. An XPS investigation of the oxidation/corrosion of melt-spun Mg. Applied Surface Science 2000; 158: 112-119.

[44] Xu Y, Huo K, Tao H, Tang G, Chu PK. Influence of aggressive ions on the degradation behavior of biomedical magnesium alloy in physiological environment. Acta Biomaterialia 2008; 4: 2008-2015.

[45] Yang L, Zhang E. Biocorrosion behavior of magnesium alloy in different simulated fluids for biomedical application. Materials Science and Engineering C 2009; 29: 1691-1696.

[46] Kirkland NT, Lespagnol J, Birbilis N, Staiger MP. A survey of bio-corrosion rates of magnesium alloys. Corrosion Science 2010; 52: 287-291.

[47] Chen J, Wang J, Han E-H, Ke W. In situ observation of pit initiation of passivated AZ91 magnesium. Corrosion Science 2009; 51: 477-484.

[48] Zong Y, Yuan G, Zhang X, Mao L, Niu J, Ding W. Comparison of biodegradable behaviors of AZ31 and Mg-Nd-Zn-Zr alloys in Hank´s physiological solution. Materials Science and Engineering B 2012; 177: 395-401.

[49] Witte F, Fischer J, Nellesen J, Crostack H-A, Kaese V, Pisch A, Beckmann F, Windhagen H. In vitro and in vivo corrosion measurements of magnesium alloys. Biomaterials 2006; 27: 1013-1018.

[50] Li ZJ, Gu XN, Lou SQ, Zheng YF. The development of binary Mg-Ca alloys for use as biodegradable materials within bone. Biomaterials 2008; 29: 1329-1344.

[51] Staigner MP, Pietak AM, Huadmai J, Dias G. Magnesium and its alloys as orthopedic biomaterials: A review. Biomaterials 2006; 27: 1728-1734.

[52] Hornberger H, Virtanen S, Boccaccini AR. Biomedical coatings on magnesium alloys - A review. Acta Biomaterialia. 2012; 8: 2442-2455.

[53] Zhong C, Liu F, Wu Y, Le J, Liu L, He M, Zhu J, Hu W. Protective diffusion coatings on magnesium alloys: A review of recent developments. Journal of Alloys and Compounds 2012; 520: 11-21.

[54] Fukumoto S, Sugahara K, Yamamoto A, Tsubakino H. Improvement of corrosion resistance and adhesion of coating layer for magnesium alloy coated with high purity magnesium. Mater. Trans. 2003; 44: 518-523.

[55] Zhang E, Xu L, Yang K. Formation by ion plating of Ti-coating on pure Mg for bio-medical applications. Scripta Materialia 2005; 53: 523-527.

[56] Cui X, Jin G, Li Q, Yang Y, Li Y, Wang F. Electroless Ni-P plating with a phytic acid pretreatment on AZ91D magnesium alloy. Materials Chemistry and Physics 2010; 121: 308-313.

[57] Hu R-G, Zhang S, Bu J-F, Lin C-J, Song G-L. Recent progress in corrosion protection of magnesium alloys by organic coatings. Progress in Organic Coatings 2012; 73: 129-141.

[58] Feil F, Fürbeth W, Schütze M. Purely inorganic coatings based on nanoparticles for magnesium alloys. Electrochimica Acta 2009; 54: 2478-2486.

[59] Boccaccini AR, Keim S, Ma R, Li Y, Zhitomirsky I. Electrophoretic deposition of bio-materials. In: J.R. Soc. Interface (ed.). 2010; 7: S581-S613.

[60] Wu G, Xu R, Feng K, Wu S, Wu Z, Sun G, Zheng G, Li G, Chu PK. Retardation of surface corrosion of biodegradable magnesium-based materials by aluminum ion implantation. Applied Surface Science 2012; 258: 7651-7657.

[61] Wang C, Li T, Yao B, Wang R, Dong C. Laser cladding of eutectic-based Ti-Ni-Al alloy coating on magnesium surface. Surface and Coatings Technology 2010; 205: 189-194.

[62] Yang J, Cui F-Z, Lee IS, Wang X. Plasma surface modification of magnesium alloy for biomedical application. Surface and Coating Technology 2010; 205: S182-S187.

[63] Zeng R, Dietzel W, Witte F, Hort N, Blawert C. Progress and challenge for magnesium alloys as biomaterials. Avanced Biomaterials 2008; 10: B3-B14.

[64] Hamu GB, Eliezer D, Wagner L. The relation between severe plastic deformation microstructure and corrosion behavior of AZ31 magnesium alloy. Journal of Alloys and Compounds 2009; 468: 222-229.

[65] Liu CL, Xin YC, Tang GY, Chu PK. Influence of heat treatment on degradation behavior of biodegradable die-cast AZ63 magnesium alloy in simulated body fluid. Materials Science and Engineering A 2007; 456: 350-357.

[66] Chang J-W, Fu P-H, Guo X-W, Peng L-M, Ding W-J. The effects of heat treatment and zirconium on the corrosion behavior of Mg-3Nd-0.2Zn-0.4Zr (wt. %) alloy. Corrosion Science 2007; 49: 2612-2627.

[67] Neil WC, Forsyth M, Howlett PC, Hutchinson CR, Hilton BRW. Corrosion of heat treated magnesium alloy ZE41. Corrosion Science 2011; 53: 3299-3308.

[68] Peng L-M, Chang J-W, Guo X-W, Atrens A, Ding W-J, Peng Y-H. Influence of heat treatment and microstructure on the corrosion of magnesium alloy Mg-10Gd-3Y-0.4Zr. Journal of Applied Electrochemistry 2009; 39: 913-920.

[69] Zhang X, Yuan G, Mao L, Niu J, Fu P, Ding W. Effects of extrusion and heat treatment on the mechanical properties and biocorrosion behaviors of a Mg-Nd-Zn-Zr alloy. Journal of the Mechanical Behavior of Biomedical Materials 2012; 7: 77-86.

[70] Orlov D, Ralston KD, Birbilis N, Estrin Y. Enhanced corrosion resistance of Mg alloy ZK60 after processing by integrated extrusion and equal channel angular pressing Acta Materialia 2011; 59: 6176-6186.

[71] Aung N, Zhou W. Effect of grain size and twins on corrosion behavior of AZ31B magnesium alloy. Corrosion Science 2010; 52: 589-594.

[72] Alvarez-Lopez M, Pereda MD, del Valle JA, Fernandez-Lorenzo M, Garcia-Alonso MC. Corrosion behavior of AZ31 magnesium alloy with different grain sizes in simulated biological fluids. Acta Biomaterialia 2010; 6: 1763-1771.

[73] Liao J, Hotta M, Yamamoto N. Corrosion behavior of fine-grained AZ31B magnesium alloy. Corrosion Science 2012; 61: 208-214.

[74] Ning Z, Cao P, Wang H, Sun J, Liu D. Effect of cooling conditions on grain size of AZ91 alloy. Journal of Materials Science and Technology 2007; 23: 645-649.

[75] Scully JR, Gebert A, Payer JH. Corrosion and related mechanical properties of bulk metallic glasses. Journal of Materials Research 2007; 22: 302-313.

[76] Wang Y, Tan MJ, Pang J, Wang Z, Jarfors AWF. In vitro corrosion behaviors of Mg67Zn28Ca5 alloy: from amorphous to crystalline. Materials Chemistry and Physics 2012; 134: 1079-1087.

[77] Pellicer E, González S, Blanquer A, Barrios L, Ibañez E, Solsona P, Suriñach S, Baró MD, Nogués C, Sort J. On the biodegradability, mechanical behavior, and cytocompatibility of amorphous Mg72Zn23Ca5 and crystalline Mg70Zn23Ca5Pd2 alloys as temporary implant materials. J. Biomed. Mater. Research A 2013; 101A: 502–517.

[78] Gu X, Zheng Y, Zhong S, Xi T, Wang J, Wang W. Corrosion of, and cellular responses to Mg-Zn-Ca bulk metallic glasses. Biomaterials 2010; 31: 1093-1103.

[79] Makar GL, Kruger J. Corrosion studies of rapidly solidified magnesium alloys. Journal of the Electrochemical Society 1990; 137: 414-421.

[80] Baliga CB, Tsakiropoulos P. Development of Corrosion Resistant Magnesium Alloys: Part II. Structure of the Corrosion Products formed on the Surfaces of Rapidly Solidified Mg-16Al Alloys. Materials Science and Technology 1993; 9: 513-519.

[81] Lunder O, Lein JE, Aune TK, Nisancioglu K. The role of magnesium aluminum (Mg17Al12) phase in the corrosion of magnesium alloy AZ91. Corrosion 1989; 45: 741–748.

[82] Corrosion resistance of aluminum and magnesium alloys: understanding, performance and testing. In: Ghali E, Revie W (ed.). Wiley series in corrosion. John Wiley & Sons; 2010.

[83] Song GL, Atrens A, Wu X, Zhang B. Corrosion behavior of AZ21, AZ501 and AZ91 in sodium chloride. Corrosion Science 1998; 40: 1769-1791.

[84] Song GL, StJohn D. Corrosion performance of magnesium alloys MEZ and AZ91. International Journal of Cast Metals Research 2000; 12: 327-334.

[85] Song GL, StJohn D. The effect of zirconium grain refinement on the corrosion behavior of magnesium-rare earth alloy MEZ. Journal of Light Metals 2002; 2: 1-16.

[86] Sun M, Wu G, Wang W, Ding W. Effect of Zr on the microstructure, mechanical properties and corrosion resistance of Mg-10Gd-3Y magnesium alloy. Materials Science and Engineering A 2009; 523: 145-151.

[87] Gu X, Zheng Y, Cheng Y, Zhong S, Xi T. In vitro corrosion and biocompatibility of binary magnesium alloys. Biomaterials 2009; 30: 484-498.

[88] Kim W-C, Kim J-G, Lee J-Y, Seok H-K. Influence of Ca on the corrosion properties of magnesium for biomaterials. Materials Letters 2008; 62: 4146-4148.

[89] Li Z, Gu X, Lou S, Zheng Y. The development of binary Mg-Ca alloys for use as biodegradable materials within bone. Biomaterials 2008; 29: 1329-1344.

[90] You BS, Park WE, Chung IS. The effect of calcium additions on the oxidation behavior in magnesium alloys. Scripta Materialia 2000; 42: 1089-1094.

[91] Kannan MB, Raman RKS. In vitro degradation and mechanical integrity of calcium-containing magnesium alloys in modified simulated body fluid. Biomaterials 2008; 29: 2306-2314.

[92] Zhang E, Yang L, Xu J, Chen H. Microstructure, mechanical properties and biocorrosion properties of Mg-Si(-Ca, Zn) alloy for biomedical application. Acta Biomaterialia 2010; 6: 1756-1762.

[93] Zhang S, Zhang X, Zhao C, Li J, Song Y, Xie C, Tao H, Zhang Y, He Y, Jiang Y, Bian Y. Research on an Mg-Zn alloy as a degradable biomaterial. Acta Materialia 2010; 6: 626-640.

[94] Quan Y, Chen Z, Gong X, Yu Z. CO_2 laser beam welding of dissimilar magnesium-based alloys. Materials Science and Engineering A 2008; 496: 45-51.

[95] Kattamis TZ. Lasers in Metallurgy. In: Mukherjee K. and Mazumder J. (ed.). The Metals Society of AIME. Warrendale, PA; 1981. p1–10.

[96] Ahmed S, Edyvean RGJ, Sellars CM, Jones H. Effect of addition of Mn, Ce, Nd and Si additions on rate of dissolution of splat quenched Mg-Al and Mg-Zn alloys in 39 % NaCl solution. Materials Science and Technology 1990; 6: 469-474.

[97] Südholz AD, Birbilis N, Bettles CJ, Gibson MA. Corrosion behavior of Mg-alloy AZ91E with atypical alloying additions. Journal of Alloys and Compounds 2009; 471: 109-115.

[98] Pietak A, Mahoney P, Dias GJ, Staigner MP. Bone-like matrix formation on magnesi-um and magnesium alloys. Journal of Materials Science: Materials in Medicine 2008; 19: 407-415.

[99] Tancret F, Osterstock F. Modelling the toughness of porous sintered glass beads with various fracture mechanisms. Philosophical Magazine 2003; 83: 125–136.

[100] Krause A, von der Höh N, Bormann D, Krause C, Bach F-W, Windhagen H, Meyer-Lindenberg A. Degradation behavior and mechanical properties of magnesium im-plants in rabbits tibiae. Journal of Materials Science 2010; 45: 624-632.

[101] Waizy H, Weizbauer A, Modrejewski C, Witte F, Windhagen H, Lucas A, Kieke M, Denkena B, Behrens P, Meyer-Lindenberg A, Bach F-W, Thorey F. In vitro corrosion of ZEK100 plates in Hank's Balanced Salt Solution. BioMedical Engineering OnLine 2012; 11: 12.

[102] Huehnerschulte TA, Angrisani N, Rittershaus D, Bormann D, Windhagen H, Meyer-Lindenberg A. In vivo corrosion of two novel magnesium alloys ZEK100 and AX30 and their mechanical suitability as biodegradable implants. Materials 2011; 4: 1144-1167.

[103] Fornell J, Concustell A, Suriñach S, Li WH, Cuadrado N, Gebert A, Baró MD, Sort J. Yielding and intrinsic plasticity of Ti-Zr-Ni-Cu-Be bulk metallic glass. Int. J. Plast. 2009; 25: 1540–1559.

Chapter 4

Emerging Trend in Natural Resource Utilization for Bioremediation of Oil — Based Drilling Wastes in Nigeria

Iheoma M. Adekunle, Augustine O. O. Igbuku,
Oke Oguns and Philip D. Shekwolo

Additional information is available at the end of the chapter

1. Introduction

1.1. Background

Nigeria is a country endowed with diverse mineral and natural resources among which is petroleum, a pivot to the national economy and sustainable development. In the past five decades, petroleum exploration and production activities have brought national economic boom but not without some aches. Acts of sabotage such as crude oil theft, pipeline bunkering and artisanal refining added to accidental spills and operational failures all combine to aggravate the oil-related aches. Oil spill into the environment, stemming from either acts of sabotage or operational failures, ultimately lead to environmental pollution with petroleum hydrocarbons [1, 2]. Petroleum mining or drilling is another factor to petroleum hydrocarbons in the environment. Most of the adverse impacts of oil spill/ petroleum hydrocarbons in the environment are experienced in the oil bearing communities, located in the Niger Delta region of the country; prominent among them being the Ogoni land pollution incidence reported by United Nations Environment Programme [1]. Petroleum exploration and production activities are strongly associated with drilling operations for oil mining. Accordingly, the extraction of petroleum resources from the earth is achieved by drilling activities. A developed drilling concept, irrespective of technological advancement, has its technical challenges, process requirements and environmental issues [3]. Drilling fluids, also referred to as drilling muds are used to enhance drilling activities via suspension of cuttings, pressure control, stabilization of exposed rocks, provision of buoyancy, cooling and lubricating.

Types of drilling fluids (muds): There are basically two categories of drilling fluids namely (i) aqueous drilling muds or water based muds (WBMs), which consist of fresh or salt water containing a weighting agent, usually barite ($BaSO_4$), clay or organic polymers and various inorganic salts, inert solids, and organic additives to modify the physical properties of the mud so that it functions optimally and (ii) non-aqueous drilling fluids (NADFs), which comprise all non-water dispersible base fluids such as oil based muds (OBMs) and synthetic based muds (SBMs) [2]. Comparative evaluation of oil based muds and water based muds shows that OBMs offer advantages over WBMs for the reasons that [3]:

- OBMs are more suitable to drill sensitive shells, allowing drilling faster than the WBMs, providing excellent shale stability

- they are more adequate to drill formulations where bottom hole temperatures exceed WBMs tolerance, especially in the presence of contaminants such as water, gases, cement, salt and temperature up to 550F

- OBMs resist formation salt leach out

- they are characterized by thin filter cakes and the friction between the pipe and wellbore is minimized, thus, reducing the risk of differential sticking and are especially suited for highly deviated and horizontal wells

- the drill of low pore pressure formations is easily accomplished, since mud weight can be maintained at a weight less than that of water (as low as 7.5 ppg)

- corrosion of pipe is controlled since oil is the external phase and coats the pipe. The oils are non-conductors and the additives are thermally stable, hence, do not form corrosive products

- bacteria do not thrive long in OBMs

- there is the possibility of using OBMs over and over again and can be stored over long periods of time since bacterial growth is suppressed

- OBM packer fluids are designed to be stable over long periods of time even when exposed to high temperature and provide long-term stable packers since additives are extremely temperature stable. Properly designed, such packer fluids can suspend weighting materials over long periods of times.

In other words, regarding shale stability, penetration rate, high temperatures, drilling salts, lubrication, low pore pressure formations, corrosion control, re-use and packer fluids, OBMs offer advantages over WBMs. It is therefore, obvious that though WBMs are more environmentally benign, they are only satisfactory for less demanding drilling of conventional vertical wells at medium depths, whereas OBMs are more suited for greater depths or in directional or horizontal drillings, which exert greater stress on drilling apparatus. As a result, OBMs are more frequently used in petroleum industries for drilling purposes. The composition of OBMs include: petroleum base fluid, weighting agent and other chemical additives.

Drill cuttings: During drilling, particles of crushed rocks produced by the grinding action of the drill bit as it penetrates the earth are referred to as drill cuttings (DC). DCs are, therefore, a mixture of rocks and particulates released from geological formulations in the drill holes made for crude oil drilling and are usually coated with the drilling fluid. Consequently, DCs are largely influenced by the chemical composition of drilling muds [2, 4].

The resultant spent OBM and drill cuttings (drilling wastes) consist of hydrocarbons, water, soils, heavy metals and water soluble salts such as chlorides and sulphates [3, 4]. Drilling wastes, which are toxic due to the presence of hydrocarbons, heavy metals and other chemical additives, if not properly treated before disposal, pose serious environmental hazards and risk to public health. Sequel to these, best practices in the management of drilling wastes cannot be over emphasized.

1.2. Health and environmental effects associated with drilling wastes

Health effects linked to drilling wastes are traceable to the basic components such as the drilling fluid and additives:

Health effects associated with drilling fluids: These health effects are attributed to the physical and chemical properties of the drilling fluids. In oil based drilling wastes, the base oil stem from petroleum stream such as crude oil, diesel (gasoil) and kerosene, which cause skin irritation. Consequently, the most commonly observed health effect associated with drilling fluids is skin irritation. Other effects include headache, nausea, eye irritation and coughing. Routes of exposure in human are dermal, inhalation, oral and some other miscellaneous routes. On exposure to drilling fluid, petroleum hydrocarbons tend to remove natural fat from the skin, which results in skin drying and cracking. These conditions allow compounds to permeate through the skin leading to irritation and dermatitis. Susceptibility to these health effects varies with individual resistance capacity and conditions of poor personal/environmental hygiene. High aromatic content fluids, especially diesel fuel contain significant levels of carcinogenic polynuclear aromatic hydrocarbons (PAHs). Diesel fuels may also be genotoxic due to high proportions of 3-7 ring PAH [2]. Skin-painting studies in mice showed that, irrespective of the level of PAH, long-term dermal exposure to diesel fuels can cause skin tumours, an effect attributed to chronic skin irritation. In humans, chronic irritation may cause small areas of the skin to thicken, eventually forming rough wart-like growths, which may become malignant. Health effects from chronic exposure to PAHs may include cataracts, kidney damage, liver damage and jaundice. Naphthalene, a specific PAH, can cause the breakdown of red blood cells, if inhaled or ingested in large amounts. Animals exposed to levels of some PAHs over long periods in laboratory studies, developed lung cancer from inhalation and stomach cancer from ingesting PAHs in food [2].

Other hydrocarbon constituents of drilling fluids are the mono-aromatics popularly referred to as BTEX (benzene, toluene, ethylbenzene and xylene). BTEX compounds are very volatile, hence, will readily evaporate in warm/hot climates of tropical regions, resulting in higher concentrations in the vapor phase. As a result, there is the possibility of exposure to human via inhalation. Exposure to high concentrations of these hydrocarbons via inhalation may result in hydrocarbon induced neurotoxicity, a non-specific effect resulting in headache, nausea, dizziness, fatigue, lack of coordination, problems with attention and memory, gait disturbances and narcosis [2].

Health effects associated with additives: In addition to the irritancy of the drilling fluid hydrocarbon constituents, several drilling fluid additives may also have irritant, corrosive or sensitizing properties. Various additives include emulsion stabilizers, pH adjusters, wetting agents, viscosifiers and fluid-loss reducing agents. For instance, calcium chloride (CaCl₂) has irritant properties and emulsifiers (such as polyamine) have been associated with sensitizing properties [3]. Specific chemical additives vary with locations.

1.2.1. Environmental effects associated with drilling wastes

Apart from health effects, environmental hazards associated with drilling wastes include land, water and air pollution [5]:

i.   **Land pollution:** Farming is the major land use system in Nigeria, especially in the Niger Delta region [1]. The most significant in this aspect of environmental pollution in Nigeria is thus farmland pollution. Consequences include alteration in soil physical, biological and chemical properties, loss of soil fertility, stunted plant growth and reduced crop productivity. These lead to reduced food security and compromised food safety.

ii.   **Aquatic pollution:** Large percentage of the oil spill gets spread over the surface of the aquatic system resulting in anaerobic environment in the water, below the surface. This leads to death of the natural flora and fauna where oxygen is the key element for their respiration; adversely affecting fishing profession [1]

iii.   **Air pollution:** volatile organics such as benzene, toluene, ethylbenzene and xylene could have elevated concentrations in the air, leading to atmospheric pollution and consequent adverse environmental and health impacts.

Oil well drilling processes generate large volumes of drill cuttings and spent mud in the country. Drilling wastes, therefore, add to hazardous petroleum waste materials released in the environments of the Niger Delta region of the country [1, 6] and the management of drilling wastes is quite tasking. An environmentally friendly technique for the management of drilling wastes is necessary in all offshore and onshore operations; from seismic surveys, drilling operations, field development and production to decommissioning. The physical and chemical properties of the drilling wastes influence their hazardous characteristics and environmental impact abilities, which in turn depend primarily on: (i) nature of impacted material, (ii) concentration of pollutant /amount of waste material after release (iii) recipient biotic community and (iv) exposure duration. Exposure that causes an immediate effect is called acute exposure while long-term exposure is called chronic exposure. Either acute or chronic exposure has negative impacts.

1.3. Contemporary treatment of drilling waste materials

Worldwide, contemporary drilling waste management options include re-use, offshore discharge, re-injection and onshore treatment and/or disposal [7]. Each treatment and or disposal option has its pros and cons as highlighted in the options (thermal technologies and bioremediation techniques) discussed.

1.3.1. Thermal treatment

As the name suggests, thermal technologies involve the use of high temperatures to reclaim hydrocarbon contaminated materials [8]. Thermal treatment is mostly used in treating organic compounds. Additional treatment may be necessary for metals and salts depending on the final fate of the wastes. Thermal treatment technologies are designed for a fixed land based installation; however, a few mobile units exist. Two commonly practiced thermal treatment technologies are thermal desorption and incineration methods.

1.3.1.1. Thermal desorption method

Thermal desorption is an environmental remediation process that uses heat to increase the volatility of contaminants by the use of a series of equipment (desorber and oxidizer) such that the hydrocarbons and water are separated or removed from the solid matrix. It is normally carried out between the temperature range of 250-650°C. At these temperatures both the lighter and heavier hydrocarbons are removed and collected or thermally oxidized by further heating to a temperature of over 850°C. The resulting solid residue has essentially no residual hydrocarbons (having been oxidized), but does concentrate salts and heavy metals. Depending upon the success of process used, recovered hydrocarbons can be used as fuel or re-used as base fluid in the drilling fluid system and the resulting solid can be disposed of in a landfill or may be used in construction (of roads and bricks). Economical, operational and environmental implications of thermal desorption include:

1. Effective removal and recovery of hydrocarbons from solids

2. Possibility of recovering base fluid and end - product could be used for brick making

3. Low potential for future liability

4. Requires short time

5. High cost of handling environmental issues

6. Large volume of wastes is required to justify the cost of operation

7. Requires tightly controlled process parameters

8. High operating temperatures can lead to safety risks

9. Requires several operators

10. Heavy metals and salts are concentrated in residual solids

11. Process water contains some emulsified oil

12. Residue ash requires further treatment before disposal

13. End product is sterile and can no longer support plant Life.

1.3.1.2. Incineration method

Incineration involves (i) heating oil based mud and drill cuttings to a higher temperature range (1200-1500°C) in direct contact with combustion gases and (ii) oxidizing the hydrocarbons [8]. Solid/ash and vapor phases are generated. The gases produced from this operation may be passed through an oxidizer, wet scrubber, and bag house before being vented to the atmosphere. Stabilization of residual materials may be required prior to disposal to prevent constituents from leaching into the environment. Incineration of drilling wastes occurs in rotary kilns, which incinerate any waste regardless of size and composition. Incineration systems are designed to destroy only organic components of waste; however, most drilling wastes are non-exclusive in their content and therefore will contain both combustible organics and non-combustible inorganic materials. By destroying the organic fraction and converting it to carbon (IV) oxide and water vapor, incineration reduces waste volume. Inorganic components of wastes fed to an incinerator cannot be destroyed, only oxidized. The major inorganic materials are chemically classified as metals. Generally, these metals will exit the combustion process as oxides of the metals that enter. Economical, operational and environmental implications of incineration are as listed:

1. Low potential for future liability

2. High cost per volume

3. Heat produced could be used for energy generation

4. High energy cost

5. Requires air pollution control equipment because of safety concerns

6. At high temperatures, salts can form acid components

7. Air emissions pose environmental concerns.

In line with best practices, for thermal technologies, there is need for proper placement of end product. Demonstration of sufficient compliance with current regulations and adequate safety measures to cater for the potential risks of exposure to high temperatures.

1.3.2. Bioremediation technique

Bioremediation technique relies on the ability of microorganisms (mostly combination of bacteria) to feed on the hydrocarbons (HCs) as substrate, converting them into carbon dioxide, water and harmless clean solids; and the ability of some of the HCs to biodegrade over time. But in most cases, the native microorganisms are often overwhelmed by the extent of the hydrocarbon contamination and thus would require external nutrients to boost (bio-stimulation) their activity and ability to take up the HCs at a faster rate. In other cases, the native microorganisms may be needing help from their kind or other species of micro-organism which are grown or inoculated (bio-augmentation) in the laboratory and then introduced in the habitat of the native micro-organisms. Bioremediation could be carried out at the site of contamination (in-situ bioremediation technique) or off the site of contamination (ex-situ bioremediation technique). Bioremediation technologies include land farming, use of bioreactors, biopiles and compost- based technologies. Economical, operational and environmental implications of conventional bioremediation technique [9, 10, 11, 12, 13, 14] include:

1. Relatively inexpensive

2. Requires simple equipments and eliminates transportation cost as drill wastes could be treated on site

3. Less capital but may be labour-intensive.

4. Low maintenance cost; being a simple technology process that requires few machines, there are few delays due to equipment down-time

5. Process is fairly flexible and can be used for most drill wastes including OBM, NADFs, previously extracted materials and newly drilled cuttings

6. Proven technology

7. Requires a considerable period of time to complete a process

8. Appropriate bacteria and nutrient selection could be a daunting task

9. In cases where bacteria are inoculated and brought on site, adaptability to their new environment may hamper their performance

10. Minimal operation hazards

11. Environmentally friendly: once the contaminants have been degraded, the microbial population reduces considerably as they have used up their food source

12. Less impact on the environment as residue from process (TPH < 1%) may require no further treatment and could be used for agricultural purposes.

Recommended best practices for bioremediation technology include ensuring (i) proper initial physical, biological and chemical characterizations to determine extent of organic and inorganic contamination, (ii) required skill and persistence for the selection of several combinations of bacteria and nutrients that can provide the desired result (iii) proper periodic tillage to provide for proper aeration that facilitates degradation of the HCs and (iv) an accurate and appropriate TPH level check in between treatment process in order to monitor progress of the remediation process. Choice of waste management options typically considers local regulations, environmental assessment, cost/benefit analysis and the composition of the drilling wastes. The Department of Petroleum Resources [15] via the Environmental Guidelines and Standards for the Petroleum Industry in Nigeria (EGSPIN) stipulated guidelines on drill cuttings discharge for inland / near-shore and offshore deep water in order to minimize the adverse impact on the surrounding environment. These requirements call for an appropriate drill cuttings treatment prior to disposal in order to meet the stipulated conditions.

1.4. Review of emerging trend in the treatment of drilling waste materials in Nigeria

There are scientific evidences showing that drilling wastes generated in the country contain toxicants that are of environmental concerns. For instance, the reports of [16] on the determination of selected physical and chemical parameters including metals concentrations in a certain drill cutting dump site in the country. Results from their study showed that oil and grease on the surface and 20 feet around the waste dump area were above the specified limit [15]. There was also lack of plant growth noticed in the study, attributed to depletion of nitrogen, phosphorus and potassium values below threshold levels for plant growth. The reports of [4] on hydrocarbon and some metal contents of drilling muds and cuttings generated during the drilling of Igbokoda onshore oil wells gave total petroleum hydrocarbon (TPH), aliphatic hydrocarbon (AH) and polycyclic aromatic hydrocarbon (PAH) as generally exceeding stipulated limits by both national and international agencies. The studies of [17] on the compositional distribution and sources of polynuclear aromatic hydrocarbons (PAHs) in Nigerian oil-based drill-cuttings, showed that the total initial PAHs concentration of the drill cuttings was 223.52 mg/kg while the initial individual PAHs concentrations ranged from 1.67 to 70.7 mg/kg, dry weight, with a 90% predominance of the combustion-specific 3-ring PAHs.

The commonly employed remediation techniques for drilling wastes in Nigeria appear to be thermal technologies. However, due to economical, operational and environmental implications of these thermal technologies; search for more acceptable techniques commenced. There is scarcity of literature on the use of natural resource materials for the remediation of drilling wastes in Nigeria. The few literature resources showed that a large percentage is still at the bench-scale platform. For instance, [18] isolated *Staphylococcus sp.* from oil-contaminated soil that was treated with 1% drilling fluid base oil (HDF-2000). Their study revealed that *Staphylococcus sp.*, is a strong primary utilizer of the base oil and has potential for application in bioremediation processes involving oil-based drilling fluids. On the other hand, the effectiveness of 2 bacterial isolates (*Bacillus subtilis* and *Pseudomonas aeruginosa*) in the restoration of oil-field drill-cuttings contaminated with polynuclear aromatic hydrocarbons was studied by [19]. In that study, a mixture of 4 kg of drill cuttings and 0.67 kg of top-soil were fed into triplicate plastic reactors labeled A1 to A3, B1 to B3, C1 to C3 and O1 to O3. These were left quiescent for 7 days under ambient conditions, followed by the addition of 20 mL working solution of pure cultures of *Bacillus* sp and *Pseudomonas* sp (each of cell density 7.6 x 10^{11} cfu/mL) to reactors A1 - A3 and B1 - B3 respectively. Another 20 mL working solution containing both cultures at cell density 1.5 x 10^{12} cfu/mL was added to reactors C1 - C3. The working solution was added to each reactor (excluding the controls, O1 - O3) every 2 weeks. Mixing and watering of the set-ups were carried out at 3 days interval under ambient temperature of 30°C for a period of 6 weeks. Results showed that the predominant 3-ring PAHs, which made up 90% w/w of the total PAHs concentration of 223.52 mg/kg, were degraded below detection and the 4-ring PAHs were reduced from 4 to 0.6% by *Pseudomonas* while *Bacillus* reduced 3 and 4-ring PAHs respectively to 0.2 and 0.8%. Their works revealed that Pseudomonas degraded 3 and 4-ring PAHs relatively better than *Bacillus*. Both strains of bacteria degraded 5 and 6-ring PAHs below detection limits. Furthermore within the 3-ring PAHs, each of the strains of bacteria reduced phenanthrene to approximately 0.2%, whereas both degraded homologues acenaphthylene, acenaphthene and fluorene as well as anthracene below detection limits. For 4-ring PAHs, *Pseudomonas* degraded fluoranthene and benzo[a]anthracene. *Bacillus* also degraded benzo[a]anthracene below detection limits. *Pseudomonas* was able to reduce pyrene and chrysene to 0.3 and 0.2% respectively; whereas *Bacillus* reduced fluoranthene, pyrene and chrysene to 0.1, 0.01 and 0.4% respectively. However, treatment with the mixed culture resulted in limited degradation of 5-ring PAHs particularly in the fourth week, which was attributed to the phenomena of co-metabolism and inhibition.

The works of [20] compared the potentials of bio-augmentation and conventional composting as bioremediation technologies for the removal of PAHs from oil-field drill-cuttings. From a mud-pit, close to a just-completed crude-oil well in the Niger Delta region of Nigeria, 4000 g of drill cuttings was obtained and homogenized with 667 g of top-soil (to serve as microbes carrier) in three separate reactors (A, B and C). The bio-augmentation of indigenous bacteria in the mix was done by adding to reactors A and B a 20-mL working solution (containing 7.6×10^{11} cfu/mL) of pure culture of *Bacillus* and *Pseudomonas*, respectively, while a 20-mL working solution (containing 1.5×10^{12} cfu/mL) of the mixed culture of *Bacillus* and *Pseudomonas* was added to reactor C. The bio-preparation was added to each reactor (excluding the control) every two weeks for six weeks. The composting experiment was conducted in a 10-litre reactor in which 4000 g of drill cuttings, 920 g of topsoil and 154 g of farmyard manure and poultry droppings were homogenized. Mixing and watering of the set-ups were carried out at 3 days interval under ambient temperature over a period of six weeks. Results showed that initial individual PAHs concentrations in the drill cuttings ranged from 1.67 to 70.7 mg/kg dry weight, with a predominance of combustion-specific 3-ring PAHs (representing 90% of a total initial PAHs. After the bioremediation exercise that lasted for 42 days, total PAHs in the drill cuttings were reduced from 223.52 to 4.25 mg/kg, representing a 98.1% reduction. Away from the use of microbial strains in the treatment of drilling wastes, a bench-scale investigation was carried out by [21] to demonstrate the efficacy of technique referred to as 'Dispersion by Chemical Reaction (DCR) technology".This particular method involved the use of hydrophobized calcium oxide (CaO) to form a dry, soil-like material that could be useful in construction works.

On the other hand, after the study on the response of four phytoplankton species in some sections of Nigeria coastal waters to crude oil in controlled ecosystem [22], that revealed the adverse impacts; a multidisciplinary environmental remediation research group (ERRG) was inaugurated with the mandate to embark on innovative, cutting-edge research and development (R & D) initiative, aimed at the development of an indigenous technology for an eco-friendly technique in the treatment of soils, sediments, sludge and drilling wastes polluted by petroleum hydrocarbons, using natural products of Nigeria origin. The goal of ERRG is to translate the technology from bench-scale to field scale and come out with on- the - shelf products that will find use for both onshore and offshore remediation works. The first phase of the R & D initiative was the exploration of the remediation potential of conventional composting technology based on the results from the works of [23]. A good start was the production of a scientifically formulated and classified compost bulk [24] that are potentially viable for environmental remediation projects [25] and able to biodegrade petroleum hydrocarbons embedded in soil and related matrices [26]. The next phase was to assess public acceptance of the principles of this technology, which culminated to the reports of [27] on population perception impact on value-added solid waste disposal in developing countries, a case study of Port Harcourt City. The feedstock utilized in product formulations in this emerging, indigenous and innovative technology is 100% biodegradable and very abundant in the Nigerian environment. Consequently, the technology has been categorized by stakeholders [27] as:

i.   eco-friendly environmental remediation technique

ii.   waste to wealth initiative

iii.   waste to resource initiative

iv.   value-added waste management option

v.   a contribution to the promotion of local material development that has the potential for:

• wealth creation

• job creation

• poverty alleviation

• sound environmental management of hydrocarbon polluted wastes from the petroleum industries.

ERRG observed that either conventional composting technology or bioremediation via utilization of pure microbial isolates/strains has limitations in terms of serving the practical needs of the petroleum industry in Nigeria with regards to meeting (i) regulatory remediation targets at close – out of project and (ii) project delivery time. Subsequently, through series of bench-scale and screen house remediation investigations, products were formulated to enhance the speed of bioremediation process using nano-scale green catalysts, a technique that matured into Compost - based Nanotechnology in Bioremediation (CNB-Tech). The research group then subjected the CNB-Tech products to different scientific evaluations in order to ascertain (i) efficiency on biodegradation of petroleum hydrocarbons in oily wastes such as crude oil impacted soils, sludge and drilling wastes (drill cuttings and oil-based mud) and (ii) environmental impacts with emphasis on soil quality. Published works on assessment and prognosis of products' impact on soil quality include:

a.   Assessing the effect of bioremediation agent from local resource materials in Nigeria on soil pH [28]

b.   Impact of bioremediation formulation from Nigeria local resource materials on moisture contents for soils contaminated with petroleum [29]

c.   Assessing and forecasting the impact of bioremediation product derived from Nigeria local raw materials on electrical conductivity of soils contaminated with petroleum products [30]

d.   Soil temperature dynamics during bioremediation of petroleum products using remediation agent from Nigerian local resource materials [31].

Other works on CNB-Tech products' evaluations including (i) effect on soil heavy metal dynamics and (ii) impact on soil microbial species population and diversity are being considered elsewhere for publication. Having recorded a huge success during the laboratory scale investigations where maximum of 4000g of sample bulk and freshly hydrocarbon contaminated soils (similar to the quantities used by other investigators) [19, 20] were treated, it became necessary to assess the efficiency of CNB-Tech products on waste materials with complex nature and higher degree of hydrocarbon pollution. This aspiration was realized in collaboration with the Remediation Department of Shell Petroleum Development Company (SPDC), Port Harcourt, Nigeria through the University Liaison Team of SPDC. Sequel to this, pilot-scale projects were commissioned to evaluate the efficiency of CNB-Tech products on the degradation of hydrocarbon compounds in the following petroleum impacted materials:

i.   Hydrocarbon polluted clay soils from Ejama-Ebubu legacy site of SPDC

ii.   Hydrocarbon polluted carbonized soil from Ejama-Ebubu legacy site of SPDC

iii.   Hydrocarbon polluted sludge from Ejama-Ebubu legacy site of SPDC

iv.   Oil-based mud and drill cuttings generated from SPDC operations.

Ejama Ebubu is one of SPDC's legacy sites of up to 42 year long pollution as at the time of study in 2011 [1]. In this chapter, the efficacy of CNB-Tech products in the biodegradation of petroleum hydrocarbons in oil-based drilling wastes (OBM-DC) is presented.

1.5. Research justification

The treatment of drilling wastes, especially OBM-DC in an environmentally sound manner is a challenging task due to the complex nature of the wastes. The most popular technique adopted for the treatment of OBM-DC, thermal desorption [15] has its accompanying environmental concerns. For instance, thermal treatment technologies are associated with prohibitive capital and operational cost implications, threatening environmental consequences in addition to high occupational hazards and generation of secondary waste stream that has to be treated at extra high cost before final disposal. Consequently, there is need for a pragmatic shift to seek alternative techniques that will address the need of the oil and gas sector in the management of drilling wastes in terms of remediation target delivery time and compliance to regulatory standards in Nigeria. Regulatory standards for close-out of remediation projects vary from one country to another and success factors of a given technology are dependent on indices such as:

a.   climatic conditions

b.   geographical characteristics of the location

c.   nature and complexity of contamination

d.   expected utility of the end-products of the remediation exercise

It then becomes evident that a successful remediation technology in one part of the globe may not necessarily be efficient in another region, pointing to the need to look inward for a more practical approach to solving the environmental challenges posed by petroleum hydrocarbon polluted waste streams in Nigeria [1]. Having run laboratory, bench- scale and screen-house remediation works using CNB-Tech products on fresh hydrocarbon contaminated soils, it became necessary to conduct pilot scale remediation works on more challenging waste streams such as weathered petroleum impacted soils, sludge, sediment, oil- based drilling mud and drill cuttings, hence this project.

1.6. Research objectives

The current study comprised three major objectives:

i.   to conduct a review on the emerging trends in the treatment and related studies for drilling wastes in Nigeria,

ii.   to assess the efficiency of an indigenous and innovative application of compost - based nanotechnology in bioremediation (CNB-Tech) in biodegradation of hydro-carbons found in oil-based mud and drill cuttings; generated by a petroleum industry in Nigeria

iii.   to investigate the beneficial utility of the remediation end-product for agricultural purpose (crop production), which is a major land use system in Nigeria.

2. Research methodology

The research methodologies employed in this study were:

i.   Literature review to provide an insight to the current and emerging trend in the treatment of drilling waste materials in the country and

ii.   Practical, ex-situ, pilot scale execution of biodegradation of hydrocarbon compounds in oil-based mud and drill cuttings generated by an oil company in Nigeria using an indigenous and innovative biotechnological (CNB-Tech) approach anchored on the use of natural resource materials of Nigeria origin.

2.1. Pilot-scale remediation of oil-based mud and cuttings using CNB-Tech method

This study was carried out during the 2010/ 2011 Sabbatical Programme of the University Liaison Team of Shell Petroleum Development Company (SPDC); in conjunction with the Remediation Department of SPDC, Port-Harcourt, Nigeria. The indigenous remediation products (CNB-Tech products) prepared from cellulosic natural resource materials and biogenic nanopolymers of Nigeria origin used for this pilot remediation study, were denoted as (i) Ecorem, (ii) Bioprimer and (iii) Biozator. The last two products are solids that are transformed to the aqueous form before use while the first product is used in the solid form.

2.1.1. Project site description

The present pilot-scale project, for the purposes of adequate monitoring and efficient execu-tion, was carried out in the Industrial Area of Shell Petroleum Development Company, Port

Harcourt, Rivers State; known as "Shell IA". The earmarked project area was a relatively isolated open green field within Shell IA and according to design, a temporary sheltered facility constructed to suit the project design was erected at the site and all necessary health and safety issues were taken into consideration. The sheltered project facility comprised of three major units:

- Remediation execution section: where actual remediation took place

- Phyto-analytical section: where effects on plant life were investigated

- Mini- chemical laboratory: where necessary onsite chemical evaluations were conducted.

2.1.2. Pilot scale remediation procedure

The batch of oil-based mud and drill cuttings (OBM-DC) used in this study was generated from SPDC's operations and supplied by one of the company's certified vendors. During the conveyance procedure for OBM-DC, chain of custody document and waste stream tracking manifest was observed. Basic highlights for CNB-Tech application mode are outlined in Figure 1. Pretreatment involved recovery of free phase base fluid and stabilization involved modification of viscosity parameter.

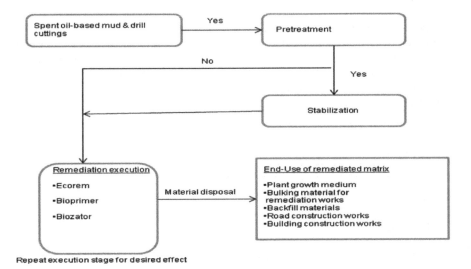

Figure 1. Application model of CNB-Tech remediation method

The biocell utilized for the remediation execution was designed by the research group, locally fabricated and lined with appropriate PVC materials. The procedures involved in the pilot remediation exercise are described as follows: A biocell of total dimension 15 m^3 was subdivided to smaller units of 3 m x 1 m x 1 m to allow for five times replication. Ecorem (a CNB-Tech product) was placed in the cells prior to loading of oil-based drilling mud and cutting

(OBM-DC) that have been previously conditioned using intervention CNB-Tech products. As the initial microbial population in OBM-DC was less than 2.0×10^3 cfu/mL, Ecorem was introduced at 10% by weight of waste materials. Using mechanical means, OBM-DC and Ecorem were homogenized and allowed to incubate for about 12 to 24 hours in order to trigger and stimulate natural microbial activities. CNB-Tech products (Bioprimer and Biozator) were then applied to saturate the contents in the biocells, which was followed by homogenization using mechanical devices. A CNB-Tech product was added to the leachate (process fluid) to immobilize inorganic constituents (especially metals) before recycling the leachate into the treatment network in such a manner that no leachate was produced as a by-product for discharge into the environment. OBM-DC that received no treatment served as control. Both controls and test units were subjected to the same environmental conditions.

System maintenance and monitoring: During remediation, the system was monitored for relevant environmental factors such as moisture content (I), pH (II), nitrogen content (III) and temperature (IV) using standard procedures of gravimetry for I, probe method via a calibrated pH meter for II, Kjedahl method for III and calibrated mercury in glass thermometer for IV. These environmental factors were maintained at the required range. Remediation lasted for 33 days: 6 days for actual treatment and 27 days for material fallow and recovery periods during which the treated materials were conditioned with a CNB-Tech product (Ecorem) for use as plant growth medium.

In order to validate the efficacy of this technology, representative composites were sent to an International Laboratory (RespirTeK Consulting Laboratory and affiliate Laboratories based in the United States of America) for physical, chemical and microbial assessments. RespirTek is ISO/EC accredited and certified. Three other laboratories that are based in Nigeria (certified by national regulatory bodies) were also involved in sample collection and analyses. Laboratories that participated in this study were:

1.   Technology Partners International Nigeria Limited, Port Harcourt - Nigeria

2.   Laser Engineering and Resources Consultants Limited, Port Harcourt- Nigeria

3.   Fugro Nigeria Limited, Port Harcourt, Nigeria

4.   RespirTek Consulting Laboratory - United States of America

2.2. Sample collection

At the end of the pilot remediation project using CNB-Tech products, treated materials were moved from the biocells and spread out on PVC impermeable membranes (each of dimension 650 cm for length and 248 cm for width), homogenized using mechanical means and air-dried with occasional homogenization of samples. The dry samples were returned into the biocells where further homogenization procedure was carried out. Sampling containers were sent by RespirTEK Consulting Laboratory, USA for their own use.

General sample collection: Using mechanical means, treated and dried samples in the cells were thoroughly homogenized for one week. In order to collect sample from a particular replicate, each replicate was subdivided into 4 equal parts; representative fractions were collected from the different parts and recombined to give a composite sample of 1kg.

BTEX sampling: Standard sampling kit for BTEX, sent by RespirTEK Consulting Laboratory, was utilized for the purpose. In this procedure, homogenized samples were collected from the cells using "Terra Core" sampling device. Using a 40 mL glass VOA vial containing appropriate preservatives and with the plunger seated in the handle, the Terra Core was pushed into freshly homogenized sample until the sample chamber was filled to the capacity of 5g. All sample particulates (debris) were removed from the outside of the Terra Core sampler and the sample plug was pushed into the mouth of the sampler. Excess soil that extended beyond the mouth of the sampler was removed. The plunger was then seated in the handle and rotated until it aligned with the slots in the body. The mouth of the sampler was placed into the 40 mL VOA vial containing the preservatives and sample extruded by pushing the plunger down. The lid was quickly placed back on the 40 mL VOA vial. It was ensured that when capping the 40 mL VOA vial, sample debris was removed from the top of the vial.

All samples were appropriately labeled and recorded in the chain of custody form before shipping to the USA laboratory by courier. Two Laboratories in Nigeria also collected samples for analyses, following standard procedures. The third laboratory in Nigeria was only involved in the analysis of materials using infrared and UV-absorption spectroscopic methods.

2.3. Physicochemical analysis and microbial assessment

Statement from quality control and quality assurance unit (QA/QC) of RespirTek Laboratory, USA showed that all analyses were conducted following procedures set forth by the ISO/IEC 17025:2005 accreditation program standards for which the laboratory holds certification. Quality assurance systems and quality control criteria were strictly followed. The following parameters were determined:

- Total petroleum hydrocarbons (TPH)

- Monoaromatic hydrocarbons: benzene, toluene, ethylbenzene and xylene (BTEX). For xylene, ortho -, meta - and para- derivatives were assessed

- PAHs: a total of 17 PAH compounds: (i) naphthalene, (ii) acenaphthylene, (iii) acenaph-thene, (iv) fluorene, (v) phenanthrene, (vi) anthracene, (vii) fluoranthene, (viii) pyrene, (ix) benzo (a) pyrene, (x) chrysene, (xi) benzo (b) fluoranthene, (xii) benzo (k)fluoranthene, (xiii) benzo (a) pyrene, (xiv) dibenz(a,b) anthracene, (xv) benzo (ghi)perylene, (xvi) 2-methyl-naphthalene and (xvii) indeno (1,2,3-cd) pyrene

- Metals: barium (Ba), calcium (Ca), copper (Cu), lead (Pb), mercury (Hg), Nickel (Ni), Sodium (Na), Potassium (K), cadmium (Cd), zinc (Zn) and arsenic (As), a metalloid

- Miscellaneous parameters: pH, salinity, nitrogen, phosphorus, total organic carbon and electrical conductivity.

- Microbial activity: assessment of 48 hr and 96 hr microbial activities of both remediation end-product and contaminated material (control) was conducted by the USA based laboratory. Total hydrocarbon utilizing bacteria as well as total microbial count were assessed by the Nigerian based laboratories.

Hydrocarbon compounds were analyzed using Gas chromatographic method, microbial assessment was carried out using heterotrophic plate count method and metals were determined using atomic absorption spectroscopic technique. All the other parameters were carried out using standard procedures such as described in [24, 25, 32]. The CNB-Tech products (Bioprimer and (Biozator) were characterized using infrared and UV-visible spectroscopic methods. The basic characteristics of Ecorem have already been reported in [24, 25] but was slightly enhanced, in this study, for case specificity.

2.4. Assessment of seed germination potential of treated samples

The remediated materials used in this evaluation were not mixed with external soil and no external fertilizer material was added to the remediated soil. Seed germination potential (SGP) of treated samples were assessed and only viable maize seedlings were used for this purpose. In a remediated material matrix (4kg material contained in an experimental plastic pot), 6 seedlings of maize were sown. This was replicated three times. All together, 18 (6 x 3) seedlings were used to evaluate this effect. Similar set- ups were also established for the untreated oil – based mud and cuttings, which served as control systems. This gave a total of 18 (6 x 3) seeds tested for germination potential for the test systems and 18 seedlings for the control media. This phase of the evaluation lasted for 7 days.

2.5. Assessment of process fluid (leachate) effect on plant growth

Adequate leachate (process fluid) management strategy was put in place as leachate generated during remediation was recycled into the remediation process. However, this evaluation was to ensure or to prove that in the event of any leachate seepage there would be reduced environmental risk. This phytotoxicity assessment was carried out using a cereal (corn: Zea mays L.,) as an indicator crop and indices of toxicity were (i) root length and (ii) plant height. Experimental systems constituted of the following set-ups, where FS is dilution factor and SF stands for farm soil:

i.   Farm soil + tap water (Code: FS + water). This served as control system for (ii) and (iii)

ii.   Farm soil + stock leachate (Code: FS + LDF-0). This served as control system for (iii)

iii.   Farm soil + diluted leachate series:

a.   Farm soil + leachate DF-1 (Code: FS + LDF-1)

b.   Farm soil + leachate DF-2 (Code: FS + LDF-2)

c.   Farm soil + leachate DF-3 (Code: FS + LDF-3)

d.   Farm soil + leachate DF-4 (Code: FS + LDF-4)

For this assessment, bulk farm soil sample, obtained from a village (K-dere, part of Ogoniland) in Rivers State, was used. Soil was sieved through a mesh and transferred at 1.5 kg per pot and designated pots were treated to 70% approximate field capacity (determined against gravity) using equal volume of appropriate fluid (water, stock leachate or diluted leachate). The systems were allowed to stabilize for 2 weeks after which viable maize seedlings were sown at 3 per pot. As the plants grew, the soil systems were treated with equal volumes of the appropriate fluid to maintain appropriate moisture level, as required by plant. Experiment lasted for 2 weeks, at the end of which the heights were recorded and plants harvested. Caution was exercised to ensure that roots were not destroyed during harvest. Root lengths were then recorded and mean values per pot calculated for each parameter.

2.6. Evaluation of beneficial utilization of end-product

Similar to the case in Section 2.4, in this evaluation, the remediated matrix was not mixed with any type of soil, neither was any external fertilizer administered. At close - out of the pilot-scale remediation project, the remediated materials were air dried, primed with one of CNB-Tech products (Ecorem) at a specified loading scheme and then utilized as a growth media. Primed end-products were transferred at 4 kg per pot of 4 liter capacity. Three indicator crops used for this project were:

- Corn (*Zea mays L.,*)

- Green leafy vegetable (Fluted pumpkin: *Telfairia Occidentalis*)

- Cassava (*Manihot esculenta Crantz*)

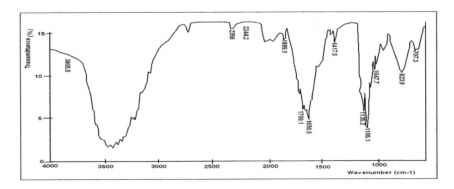

Figure 2. Infrared spectrum of Bioprimer, a CNB-Tech remediation product

The crops were used because they are commonly grown and consumed in the Niger Delta region of the country. Due to time constraint, duration of investigation varied for the crops, the longest being up to 130 days for green leafy vegetable (Fluted pumpkin: *Telfairia Occidentalis*) while corn (*Zea mays L.,*) and cassava (*Manihot esculenta Crantz*) were grown for 2 and 3 weeks respectively. Untreated OBM-DC served as a control and farm soil served as a second control.

2.7. Statistical analysis

Data generated in this study were subjected to statistical evaluations using SPSS software for Windows, version 17.0. Descriptive statistics were applied to evaluate mean and standard deviation. Paired sample T-Test and One-way analysis of variance (ANOVA) were applied to identify significant variations among treatments as appropriate. Pearson correlation was used to ascertain significant relationships.

3. Results

3.1. Typical infrared spectra of two CNB- Tech remediation products

The infrared absorption spectra of two CNB-Tech products (Bioprimer and Biozator) utilized in this pilot scale study are presented in Figures 2 and 3. Both spectra showed absorption peaks in the region of 4000 to 600 cm^{-1}.

Major information from the infrared spectra were: strong, broad absorption band of oxygen-hydrogen (O-H) of an alcohol (aryl/aliphatic) and N-H absorption bonds around 3500 - 3300 cm^{-1}; carbon-oxygen double bond (C=O) absorption band found around 1750 – 1500cm^{-1} This could be carbonyls of ester (RCOOR), aldehyde (RCHO), ketone (RCOR) and acid (RCOOH). C-N bond of nitrogenous matter falls in the end of the range; C-O bond around 1200 – 1000 cm^{-1} and of carbon-hydrogen (C-H) bond for aromatic moieties found below 1000cm^{-1} [33].

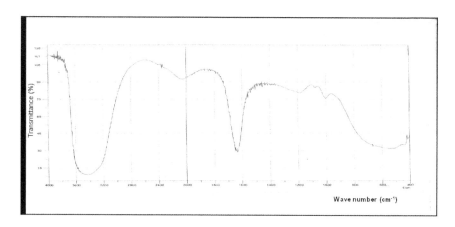

Figure 3. Infrared spectrum of Biozator, a CNB-Tech remediation product

3.2. Initial characteristics of the drilling wastes

The results presented in this paper were largely those obtained from the International laboratory. Table 1 contains the initial characteristics of the drilling wastes (oil-based mud and cuttings).

S/N	Parameter	Concentration
Inorganics		
1.	Arsenic (mg/kg)	6.69
*2.	Cadmium	Not determined
3.	Barium(mg/kg)	765
4.	Calcium(mg/kg)	87300
5.	Copper(mg/kg)	35.90
6.	Lead(mg/kg)	161
7.	Mercury(mg/kg)	0.036
8.	Nickel(mg/kg)	12.3
9.	Sodium(mg/kg)	493
10.	Potassium(mg/kg)	1930
11.	Zinc(mg/kg)	144
12.	TKN (%)	0.0357
13.	Phosphorus (%)	0.0291
*14.	pH	10.2
*15.	Electrical conductivity (mSm⁻¹)	Not determined
16	Total organic carbon (%)	Not determined
17..	Salinity (mg/kg)	4300
BTEX compounds		
1.	Benzene	0.0198
2.	Ethylbenzene	0.827
3.	m- and p-xylene	0.532
4.	o-xylene	0.924
5.	toluene	1.910
PAH Compounds		
1.	Naphthalene(mg/kg)	1.94
2.	Acenaphthylene(mg/kg)	BDL
3.	Acenaphthene(mg/kg)	BDL

S/N	Parameter	Concentration
4.	Fluorene(mg/kg)	2.54
5.	Phenanthrene(mg/kg)	0.78
6.	Anthracene(mg/kg)	BDL
7.	Fluoranthene(mg/kg)	BDL
8.	Pyrene(mg/kg)	BDL
9.	Benzo (a) anthracene(mg/kg)	BDL
10.	Chrysene(mg/kg)	BDL
11.	Benzo(b)fluoranthene(mg/kg)	BDL
12.	Benzo (k)fluoranthene(mg/kg)	BDL
13.	Benzo(a)pyrene(mg/kg)	BDL
14.	Dibenz(a,h)anthracene(mg/kg)	BDL
15.	Benzo(g,h)perylene(mg/kg)	BDL
16.	2-methylnapthalene(mg/kg)	5.39
17.	Indeno(1,23-cd)pyrene(mg/kg)	BDL
	Total PAH(mg/kg)	10.65
Total petroleum hydrocarbon		
1.	TPH (mg/kg)	79 200

*Parameters not determined by the USA laboratory but quantified by Nigerian based laboratories

Table 1. Initial characteristics of the oil -based drilling mud and cuttings used in this pilot scale study

Results indicated the presence of inorganic constituents and organics (hydrocarbons compounds). Regarding inorganics, soft metal contents increased in the order: Na (493 mg/kg) <
K (1930 mg/Kg) < Ca (87, 300 mg/kg). The elemental ratios were 177 for Ca/Na, 45 for Ca/K
and 4 for K/Na. Heavy metal concentrations increased in the order: Hg < As < Ni < Zn < Cu <
Pb < Ba. In terms of hydrocarbon contents, total concentrations of polynuclear aromatic
hydrocarbon (PAH) compounds was 10.65 mg/kg with concentrations of the individual
components (Figure 4) increasing as phenanthrene (0.78 mg/Kg: 7%) < naphthalene (1.94 mg/
kg; 18%) < fluorene (2.54mg/kg; 24%) < 2-methylnapthalene (5.39 mg/kg; 51%). Results on
monoaromatics (BTEX), shown in Figure 5, gave a total concentration of 4.213 mg/kg out of
which toluene constituted the highest fraction (45.34%), followed by xylene (34.56%), ethylbenzene (19.63%) and benzene (0.47%). Total xylene concentration was 1.456 mg/kg out of
which ortho-xylene constituted 63.46% while meta- and para-xylenes gave 36.54% of the total
(1.456 mg/kg).

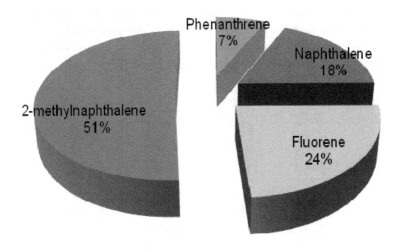

Figure 4. Percentage distribution of individual components of PAH relative to the total concentration

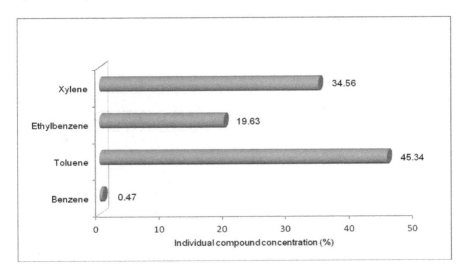

Figure 5. Percentage distribution of individual components relative to the total BTEX concentration

3.3. Results on petroleum hydrocarbon degradation

By application of CNB-Tech products, the initial TPH concentration of 79, 200 mg/kg decreased to 1888.67 ±161. 20 mg/kg. The difference in these two values was a mean TPH concentration of 77 311.33 ± 161.20 mg/kg. This difference corresponds to the total concentration of hydro-carbon compounds degraded or destroyed by the applied treatment. The initial concentration (79, 200 mg/kg) and the degraded fractions (in replicates of three) are presented in Figure 6. Specifically, results on hydrocarbon degradation (Figure 7) revealed 98% degradation for TPH, 100% degradation for BTEX and 100% degradation for PAH. Reduction in TPH level by 99% was obtained by the Nigerian laboratories.

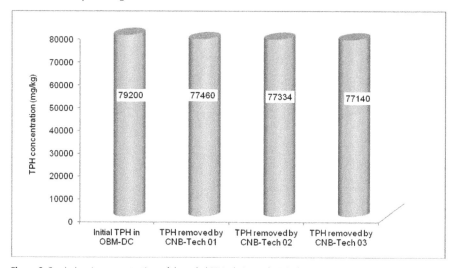

Figure 6. Graph showing concentrations of degraded TPH relative to the initial concentration

S/N	Parameter	Remarks for contaminated medium	Remarks for remediated medium
1.	Appearance	Viscous, pasty and solid interfaced in oil suspension	Transformed to non-viscous, non-sticky crumby humus soil appearance
2.	Color	Light brown	Treated matrix had characteristic dark color of humus soil
3.	Odor	Presence of strong hydrocarbon odor	Complete disappearance of hydrocarbon odor in all the treated media and all treated samples exhibited clean earthy smell
4.	Sheen test	Strong oil sheen in water suspension	Complete disappearance of oil sheen in water suspension

Table 2. Qualitative results for the remediated media

Results on qualitative assessments of the untreated OBM-DC and remediated material in terms of appearance, odor, color and sheen test are contained in Table 2 and Figure 8 depicts the materials' appearances before and after remediation.

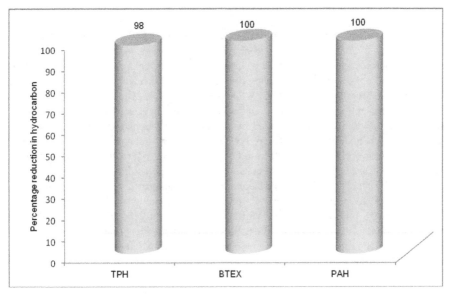

Figure 7. Percentage degradation of hydrocarbon compounds in the drilling wastes by applied CNB-Tech products

Figure 8. Photographs showing the materials before and after bioremediation by the application of CNB-Tech products

3.4. Results on inorganic constituents of the CNB -Tech treated materials

Descriptive statistics of selected inorganic constituents found in the treated media are presented in Table 3. Changes in their concentrations relative to the initial values are presented in Figure 9. For instance, the initial pH value was reduced to 7.90 from 10.20, corresponding to 23% reduction. Likewise, the following reductions were obtained: 62% for Ca, 46% for As, 44% for Cu, 70% for Pb, 100% for Hg, 57% for Ni and 37% for Zn. The concentrations of some elements such as nitrogen, phosphorus and potassium were elevated. The nitrogen-phosphorus-potassium (NPK) status, as affected by treatment, is presented in Figure 10. Nigerian laboratories obtained the same trend for NPK status. Based on the results from USA, CNB-Tech remediation option applied in this study raised the nitrogen level from 0.036% to 0.096%, raised phosphorus level from 0.0291% to 0.312%, increased potassium by 1.4 fold (Figure 10) and sodium by 3 folds. The USA based laboratory did not analyze for total organic carbon and electrical conductivity but the Nigerian based laboratory did and recorded electrical conductivity in the range of 1956 to 2063 mSm^{-1} with a mean value of 2003 ± 54 mSm^{-1} before treatment. After remediation, the electrical conductivity of the end products ranged from 594 to 696 mSm^{-1} and a mean value of 640± 52 mSm$^{-1.}$ From the mean values, there was a 68% reduction in electrical conductivity.

S/N	Element	Minimum	Maximum	Mean	Standard error	Standard deviation	Sample population
1.	pH	7.70	8.20	7.90	0.15	0.26	3
2.	Nitrogen (%)	0.070	0.130	0.096	0.016	0.028	3
3.	Phosphorus (%)	0.280	0.360	0.312	0.026	0.046	3
4.	Potassium (%)	0.50	0.77	0.61	0.08	0.14	3
5.	Copper (mg/kg)	18.10	21.70	20.10	1.06	1.83	3
6.	Zinc (mg/kg)	79.30	110	92.67	9.08	15.73	3
7.	Nickel (mg/kg)	3.99	7.05	5.29	0.92	1.59	3
8.	Calcium (mg/kg)	28900	39200	33466	3030	5248	3
9.	Arsenic (mg/kg)	2.50	4.85	3.59	0.68	1.18	3
10.	Lead (mg/kg)	5.87	54.80	27.06	14.50	25.12	3

Table 3. Concentrations of some inorganic parameters in the treated materials

Total organic carbon ranged from 2.95 to 3.06% with a mean of 2.99± 0.06% before remediation and increased to 3.84 to 3.93% with a mean of 3.88 ± 0.05%; corresponding to an increase by 23%. Before remediation, Cd concentration varied from 6.70 to 7.60 mg/kg, with a mean value of 7.03± 0.49 mg/kg. After treatment, the metal concentration ranged from 0 to 1.80 mg/kg with an average of 1.05 ± 0.94 mg/kg. By the two mean values, cadmium level was reduced by 85% due to applied CNB-Tech products.

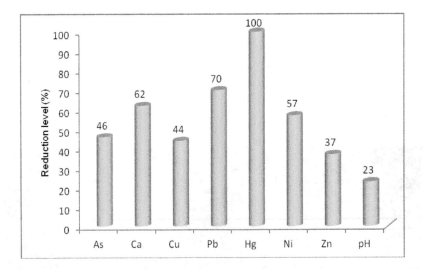

Figure 9. Reductions in some inorganic constituents of the drilling materials treated by CNB-Tech

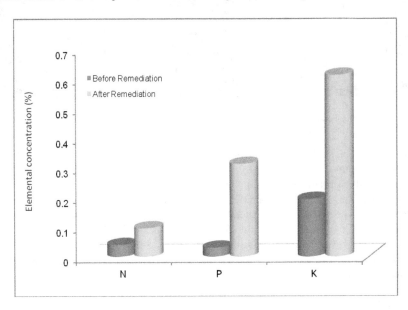

Figure 10. Nitrogen-phosphorus-potassium status before and after treatment as obtained by the USA based laboratory

3.5. Results on microbial activity

The digital photographs of heterotrophic plate count results are shown in Figure 11. Microbial activities assessed on the untreated and treated samples revealed that the contaminated oil-based mud and cuttings (no. 1 in Figure 11), contained some indigenous microorganisms of up to 1.9×10^3 (cfu/mL) while the CNB-Tech remediated samples recorded up to a maximum of 3.15×10^7 cfu/mL. An illustration of microbial enumeration for 48-hr and 96 hr counts are presented in Figure 12.

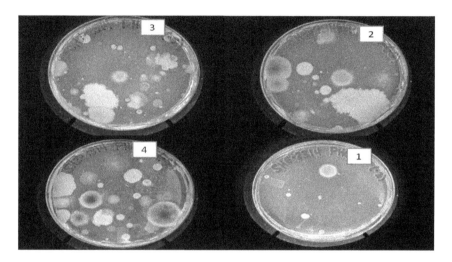

Figure 11. Heterotrophic plate count digital photographs for untreated OBM-DC (1) (before remediation) and replicates (2, 3, 4), after remediation using CNB-Tech method

At 48 hr microbial activity assessment, maximum total microbial population of 1.9×10^3 cfu/mL was obtained for untreated OBM-DC and in the materials remediated by the application of CNB-Tech products, it was 1.45×10^7 cfu/mL. These two values were significantly different at $p \leq 0.05$. At 96 hr microbial activity assessment, a total microbial population of 2.4×10^3 cfu/mL was obtained for untreated OBM-DC and 3.15×10^7 cfu/mL for the remediated matrices. Results showed that within 48 hours, the microbial activity of the remediated matrices excelled over the untreated by over 7,000 folds and at 96 hours, it excelled by over 13, 000 folds, indicating rapid multiplication of microbial activity by CNB-Tech products which also increased with time.

3.6. Results on phytotoxicity assessment of remediated samples

3.6.1. Toxicity on seed germination potential

The contaminated OBM-DC did not allow the germination of maize seedlings. Out of the sown 18 seedlings, none germinated. The untreated OBM-DC therefore, gave 100% toxici-

ty to seed germination potential (SGP) of maize. On the contrary, all the 18 maize seedlings sown in the CNB-Tech remediated matrices germinated (Figure 13). Hence, resulting
in 100% positive effect on SGP, indicating that the treated matrices exhibited 0% toxicity
to seed germination.

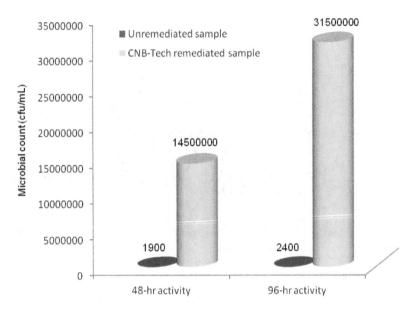

Figure 12. Microbial activity at 48 –hr and 96-hr counts for untreated oil-based drilling wastes and CNB-Tech remediated samples

Figure 13. Germinated maize seedlings growing in treated media with picture taken on day 4 of growth

3.7. Results on beneficial use of remediation end product

Figure 14, shows a cross-section of the treated materials (during recovery period) being aerated in preparation for use as plant growth media.

Figure 14. A cross section of project technical staff preparing the treated drilling wastes (OBM-DC) for use as plant growth media

During the recovery phase of the remediated end-product, treated materials were allowed to lie fallow in order to establish natural processes as a sign of wellbeing and restoration. In this project, after the fallow period, early indications of material restoration were:

- spontaneous vegetative growth,

- the presence of larva within the spontaneously grown green vegetation,

- butterflies and small birds perching on the surface of the material, which could not take place before treatment

Remediated materials supported the growth of fluted pumpkin (*Telfairia occidentalis*). A cross-section of the green leafy vegetable at over 100 days of growth and that of cassava, at one week of growth, growing in the treated materials are shown in Figure 15. Narrowing to the height of *Telfairia occidentalis*, the mean height for crops grown in the untreated OBM-DC was 0 cm as there was complete inhibition to both germination and growth. The mean height for crops grown in CNB-Tech remediated media was 217± 25 cm, a value higher than the mean height (187± 40 cm) of the vegetable crops grown in farm soil collected from the region. The difference in the two mean values was significant at $p = 0.14$. Correlation for the heights of the vegetables grown in the treated media and those grown in the farm soil gave a coefficient of 0.95 ($p = 0.204$).

Figure 15. Remediated drilling wastes as plant growth medium for Fluted pumpkin (*Telfairia occidentalis*) and cassava (*Manihot esculenta Crantz*)

3.8. Results on the impact of remediation leachate on plant life

Comparative evaluations of control system (soil treated with water only), stock leachate system (soil treated with leachate without any form of dilution) and systems treated with serial dilutions of the leachate (soil treated with leachate diluted with water by factors 1, 2, 3 and 4) are presented in Table 4.

S/N	System Code	Leachate effect of on vegetative growth relative to control (%)		Effect of serial dilution on plant using stock (undiluted leachate) as reference (%)	
		Height	Root length	Height	Root length
1.	FS+ Water (Control)	Reference	Reference	Not applicable	Not applicable
2.	FS + DF-0	-1.50	-23.45	Reference	Reference
3.	FS + DF-1	32.60	1.12	34.62	32.20
4.	FS + DF-2	45.01	16.37	42.22	50.02
5.	FS + DF-3	66.86	21.37	69.41	58.55
6.	FS + DF- 4	75.39	24.51	78.07	62.66

Negative sign stands for decrease. The other positive values stand for increase, FS = farm soil and DF = dilution factor

Table 4. Impact of leachate generated at the close-out of project on the root length and height of maize

Pictorial and graphical representations of leachate impact on plant height and root length are presented in Figures 16 and 17. Relative to the control system (soil treated with water only), leachate diluted with water by a factor of 4 improved plant height by 75.39% and root length by 24.51%. Figures16 and 17 gave all the systems at a glance, relating the control (FS + Water), system SF+LDF-0 (DF-0) and serial dilutions (DF-1 = FS+ LDF-1, DF-2 = FS+ LDF-2, DF-3 = FS + LDF-3 and DF-4 = FS + LDF - 4) for plant height and root length. Evaluating the effect of leachate dilution relative to the stock (undiluted) leachate, a 4-fold dilution excelled over the stock by 78.0% for plant height and 62.66% for root length. The relationships between plant height or root length and dilution factors are given in Figure 18. Pearson correlations gave strong coefficients: plant height versus dilution factor, r = 0.979 (p = 0.004), root length versus dilution factor, r = 0.932 (p = 0.021) and plant height versus root length, r = 0.972 (p = 0.006). From the results, plant vegetative growth increased with increasing dilution of leachate.

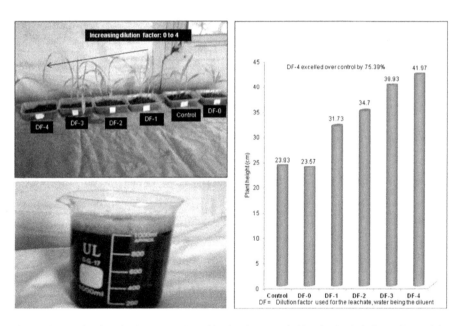

Figure 16. Pictorial and graphical representations of leachate impact on height of maize, including a picture of the stock leachate contained in a beaker

4. Discussion

The type of inorganic constituents and hydrocarbons found in the drilling wasting used in this study were consistent with the reports of [4, 17] but varied in concentrations. This confirms that the OBM-DC used in this study was toxic [2]. The remediation products of CNB-Tech

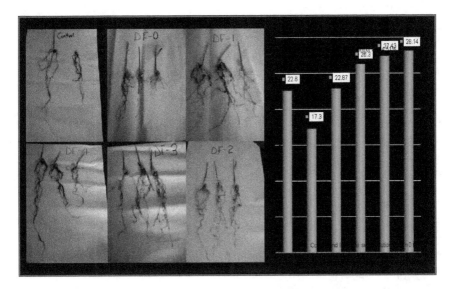

Figure 17. Pictorial and graphical representations of leachate impact on root length of maize

series used in this study demonstrated a high (98 to 100%) degradation potential for the different constituents of hydrocarbon compounds found in the drilling wastes, within a short period of 6 days. This excellent performance was attributed to the chemistry, nature and operation mechanisms of the CNB-Tech formulations.

An infrared spectrum is primarily used to identify functional groups present in a molecular fragment [33]. The infrared spectra obtained for CNB-Tech products (Biozator and Bioprimer) revealed enrichment of the molecular structure of the two products with oxo- groups, indicating oxidizing functionality. The presence of C-H of aromatic nature and the O-H stretching absorption indicate the presence of both hydrophobic and hydrophilic properties, respectively, in their molecular fragments. By implication, the remediation products are naturally endowed with:

• oxidizing ability

• polar (hydrophilic: water loving) molecular fragment

• non-polar fragment (hydrophobic: water insoluble, oil soluble) molecular fragment.

These natural endowments permit the dissolution of the products' active ingredients (solids) in water, making water the carrier medium for CNB-Tech liquid formulations. Consequently, Biozator and bioprimer are water based technical grade products. By the mentioned characteristics, the two products perform reduction and oxidation (Redox) reaction mechanisms, resulting in the degradation/ destruction of hydrocarbons compounds, without recombination

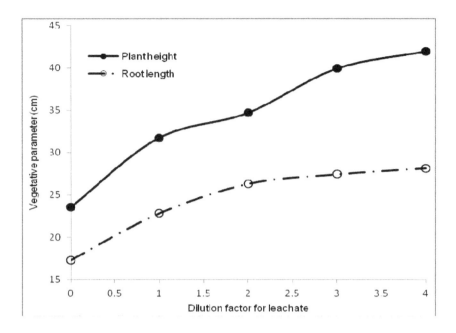

Figure 18. Relationship between plant vegetative growth and serial dilution of process fluid (leachate) generated during the remediation project

to form new hydrocarbons. These absorption peaks in the infrared spectra further reveal that CNB – Tech products are natural hydrocarbon biodegradation catalysts for the following reasons:

- enhaced water solubility of hydrocarbons via sorption, hydrolysis and oxidation mechanisms
- enhanced bioavailability of hydrocarbon pollutants for microbial degradation
- increased supply of oxygen [O] molecules required for enhanced reduction –oxidation reactions in the hydrocarbon degradation process.
- surfactant property
- emulsification of hydrocarbons

The combined actions of hydrophobic molecular fragment, hydrolysis, oxidation and surfactant property of CNB-Tech products render hydrocarbons more water soluble and subsequently more available for biodegradation. Bioprimer and Biozator also emulsify hydrocarbons into droplets that can be easily assimilated by microorganisms. By these properties, the products reduce oil-water surface tension; enhance water solubility of petroleum hydrocarbons thereby enhancing the bioavailability of the contaminants (hydrocarbons) to microorganisms for both extracellular and intracellular decompositions. The two products are 100% biodegradable. The third CNB-Tech product used in this study (Ecorem: a black amorphous solid material, also 100% biodegradable) contains major and minor plant nutrient elements and via hydro-activation, naturally generates mixed consortia of microorganisms, which multiplies with time to facilitate the destruction of hydrocarbons. No engineered microorganism or externally imported microorganism was used in this study. This technology, therefore, saves time and eliminates the daunting task of isolating pure microbial strains and associated adaptability challenges linked with conventional bioremediation techniques [7, 8, 18, 19, 20].

The microorganisms from Ecorem product perform the following functions:

• extracellular decomposition in which the naturally produced microorganisms secrete enzymes to breakdown large organic compounds (such as hydrocarbons) into smaller forms for easier absorption into the micro-organisms. Once the smaller compounds have been absorbed by the microorganisms, intracellular decomposition takes place

• increased microbial activity facilitated by Ecorem, results in thermophilic temperature modulations in the range of 55 to 60°C, a process that accelerates degradation of hydrocarbons, especially polynuclear aromatic aromatic hydrocarbons (PAHs).Thermophilic temperature modulations also controls thermo-sensitive pathogen to crops animals and man; killing off weeds and seeds that will be detrimental to land use of end products.

By the above described mechanisms, the CNB-Tech products were able to biodegrade petroleum hydrocarbon compounds with high efficiency (98% degradation for TPH and 100% degradation for PAHs and BTEX) within a short period of time of 6 days, relative to previous works on bioremediation. For instance, in a study of in-situ bioremediation of oily sludge via biostimultaion of indigenous microbes, conducted by [34], through the addition of manure at the Shengli oilfield in Northern China for 360 days, 58.2% reduction in TPH was achieved in test plots and 15.5% reduction in control plot. By treating 2 kg of drill cuttings with initial TPH of 806.36 mg/kg for 56 days under the conditions of composting of spent oyster mushroom (*P.ostreatus*) substrate, [35] recorded overall degradation of PAHs in the range of 80.25 to 92.38%. In this present study, OBM-DC used had initial TPH of 79, 200 mg/kg and was degraded by 98% within the stated short period of 6 days. In a field trial biopile composting method [36] for drilling mud polluted sites in the Southeast of Mexico with comparable TPH level of 99 300 ± 23000 mg/kg, after 180 days, TPH concentrations decreased from 99 300 ± 23000 mg/Kg to 5500 ± 700 mg/kg, corresponding to 94% degradation for amended biopile and to 22900 ±7800 mg/kg, representing 77% decrease for unamended biopile. The mean residual value of TPH (5500 ± 700 mg/kg) left in the treated matrix in their study was higher than the mean residual value (1888± 161 mg/kg) obtained in this present study.

By conducting an investigation on two bioremediation technologies (bioremediation by augmentation and conventional composting using crude manure and straw) as treatment options for oily sludge and oil polluted soil in China [12] in which the total hydrocarbon content (THC) varied from 327.7 to 371.2 g/kg (327700 to 371200 mg/kg) for dry sludge and 151.0 g/kg (151000 mg/kg) for soil for a period of 56 days; after three times of bio-preparation application, THC decreased by 46 to 53% in the oily sludge and soil. The results (98 -100% degradation) obtained from this present study was from only one dose application of CNB-Tech products. Repeated application of CNB-Tech products by two to three dose applications will achieve 100% degradation of TPH. In another instance, a 5- month field scale bioremediation of sludge matrix via the utilization of organic matter such as bark chips via conventional composting, mineral oil (equivalent to total hydrocarbons) decreased from 2400 to 700 mg/kg (70% decrease) for sludge matrix and from 700 to 200 mg/kg, corresponding to 71% decrease [14]. In treating oil sludge using composting technology in semiarid conditions for 3 months, hydrocarbons were reduced from 250 to 300g/kg (250000 to 300 000 m/kg) by 60% against reduction by 32% recorded in the control [37]. The treatment applied by [37] and consequent reduction of 60% implies that the residual hydrocarbons in the treated samples would be between 100 000 and 180 000 mg/kg unlike the results obtained in this present study that gave residual hydrocarbon of 1888.67 ±161.20 mg/kg. In a study carried out by [38], sand samples contaminated with oil spill were collected from Pensacola beach (Gulf of Mexico) and tested to isolate fungal diversity associated with beach sands and investigate the ability of isolated fungi for crude oil biodegradation. From their results, 4.7 to 7.9% biodegradation was recorded.

Elsewhere in India, Abu Dhabi and Kuwait [39], bioremediation technology was applied in field-scale degradation of hydrocarbons in different oil wastes for a period of 12 months. Table 5 illustrates different reductions in total petroleum hydrocarbons obtained in these field case studies. TPH reductions in drilling wastes were obtained in the range of 90.85 to 95.48% with residual TPH in treated samples in the range of 2600 to 10 900 mg/kg (0.26 to 1.09%).

Name of the oil Installation / type of oily waste	Quantity of oily waste (cubic meter)	Number of batches	TPH Content (%) in oily waste before and after bioremediation		% Reduction in TPH	Residual TPH in treated material (%)
			Before	After		
Abu Dhabi National Oil Company (ADNOC), Abu Dhabi / Oil contaminated drill cuttings	200	1	17.26	0.98	94.32	0.98
BG Exploration and Production India Limited (BGEPIL), India / Oil based mud (OBM)	2,428	3	5.75 – 6.23	0.26 - 0.57	95.48-90.85	0.26 – 0.57
Bharat Petroleum Corporation Limited (BPCL), India / Oily sludge	5,000	1	19. 30 – 26.5	0.26 - 0.57	98.65-97.85	0.26 -0.57

Name of the oil Installation / type of oily waste	Quantity of oily waste (cubic meter)	Number of batches	TPH Content (%) in oily waste before and after bioremediation		% Reduction in TPH	Residual TPH in treated material (%)
Cairn Energy Pty. India Limited, India / Oil contaminated drill cuttings	567	2	14.93 – 18.81	0.82 – 1.09	94.51-94.21	1.09
Chennai Petroleum Corporation Limited (CPCL), India / Oily sludge	4,444	2	26.12	0.89	96.59	0.89
Hindustan Petroleum Corporation Limited (HPCL), India / Oily sludge	5,010	3	16.70 – 52.81	0.90 – 1.60	94.61-96.97	0.90-1.60
Indian Oil Corporation Limited (IOCL) Refineries in India / Oily sludge (acidic + non acidic)	75,412	48	9.6 – 38.4	0.37 – 0.95	96.15-97.53	0.37-0.95
Kuwait Oil Company (KOC), Kuwait / Oil contaminated soil	778	1	4.6 – 12.75	0.09 – 0.10	98.04-99.21	0.09-0.10
Mangalore Refinery and Petrochemicals Limited (MRPL), India. / Oily sludge	2,222	2	8.35 – 19.86	0.84 – 0.97	89.84-95.12	0.84-0.97
Oil and Natural Gas Corporation Limited (ONGC) installations in India / Oily sludge & oil contaminated soil	95,499	145	12.0 – 51.5	0.5 – 1.2	95.83-97.67	0.50--1.20
Oil India Limited (OIL) , Assam / Oily sludge & oil contaminated soil	15,921	14	21.6 – 37.7	0.49 – 0.53	97.73-98.59	0.49-0.53

Name of the oil Installation / type of oily waste	Quantity of oily waste (cubic meter)	Number of batches	TPH Content (%) in oily waste before and after bioremediation		% Reduction in TPH	Residual TPH in treated material (%)
Reliance Energy Limited (RIL), India / Oily sludge	611	2	19.15	0.5	97.39	0.50

Table 5. Reductions in TPH levels obtained in field case studies of different types of petroleum impacted wastes (soils, drill cuttings and oil-based mud) in Abu Dhabi, Kuwait and India [39].

The residual TPH level (1888.67 ± 161.20 mg/kg) obtained in this present study was below the Environmental Guidelines and standards for the Petroleum Industry in Nigeria (EGASPIN) intervention value for mineral oil (petroleum hydrocarbon) of 5000 mg/kg [15]. By repeated application of CNB-Tech products, it is possible to meet a very strict regulatory standard for residual TPH level of less than 50 mg/kg. The changes in metal concentrations found in this study were attributed to (i) immobilization via chelate formation (ii) preferential supplementation of trace plant nutrient elements using the three products, (iii) natural electrochemical process whereby the positively or negatively charged organic molecules (generated during the natural transformation process occurring when the products were in use) bond with their counterparts in organic matter. These processes include oxidation, methylation, hydroxylation, carboxylation, coupling and polymerization [40] thereby enhancing bioavailability of the metals to microorganisms that utilize the organic matter supplied by the CNB –Tech products as energy source.

Microbial population found in a typical tropical soil under Nigerian climate is in the neighborhood of 8.19 x 10⁶ cfu/mL [41]. Relative to this value, the population found in the contaminated OBM-DC (1.9 to 2.4 x 10³ cfu/mL) showed suppressed microbial population, attributed to strong hydrocarbon (TPH level of 79, 200mg/kg) pollution. This is in agreement with the reports of [3]. The microbial population (1.45 to 3.15 x 10⁷ cfu/mL) found in treated samples revealed restoration of soil microbial population using CNB-Tech products. It excelled over the value recorded in polluted material by over 7000 folds and higher than the value reported by [34], where TPH degraders and PAH degraders increased by one to two orders of magnitude via the addition of manure. Furthermore, the use of CNB-Tech products modified the pH value of the drilling wastes, transforming it from strongly alkaline (pH of 10) medium to pH of 7.90 medium; comparable to the 7.3±0.1 obtained by [34] for bioremediated soils. The very high pH of the untreated drilling waste materials could be attributed to some of the additives in the drilling fluid. Drilling fluids contain an internal phase of brine such as calcium salts [3]. This was confirmed by the high content of Ca (87 300 mg/kg) obtained in this study for the untreated material. One dose application of CNB-Tech products reduced this concentration by up to 62%, repeated dose application would definitely bring Ca level to any desired value.

Observations made during the recovery /fallow period were signs of drastic positive change in toxicity conditions, implying reduced toxicity. Reduction of soil toxicity by bioremediation, evidenced by increase in EC50 of the soil was reported by [34]. In this study, bioremediation using CNB-Tech products reduced toxicity in treated materials relative to untreated OBM-DC, evidenced by 100% positive effect on seedling germination potential and improved crop vegetative growth. Reduced material toxicity also explains the increased microbial activity of the treated matrices in comparison to the untreated drilling wastes, obtained in this study. The agricultural potential for the remediation end-products was also manifested by:

- increased microbial activities

- increased nitrogen-phosphorus-potassium (NPK) status

- increased soil crumby nature as against very viscous and pasty characteristics of untreated drilling wastes.

These nutrient elements (NPK) enhance microbial growth, microbial population, microbial activity and consequently increase soil fertility [41]. By these, CNB-Tech products could overcome the extreme phytotoxicity [100% toxicity to seedling germination potential of maize and 100% inhibition to vegetative growth for three different types of plant (maize, fluted pumpkin and cassava)], caused by the untreated drilling waste. CNB-Tech products trans-formed oil-based drilling mud/cuttings to arable soil; capable of supporting seed germination and plant growth; excelling the performance of a control (farm soil apparently not impacted by drilling waste or crude oil) by 14%.

Electrical conductivity, a measure of dissolved ions in solution, is influenced by several soil physical and chemical properties such as salinity, saturation percentage, water content, bulk density, organic matter content, temperature and cation exchange capacity of the soil matrix. Impact of these influencing factors must be reflected in interpreting electrical conductivity effect on plant growth. Generally, elevated electrical conductivity and high salinity levels in agricultural soils may result in reduced plant growth and productivity or in extreme cases, the elimination of crops and native vegetation [42]. The reduction of electrical conductivity by 68% is a positive development because it demonstrates that the products could also modify the salinity of the material. In situations of very high initial electrical conductivity, there is a step-down CNB-Tech product as was carried out in this study and in situations of very low electrical conductivity, there is also a step-up CNB-Tech product as reported in a previous publication [30].Results in this present study on excellent growth of crops planted in the remediated matrices were indicators of acceptable soil salinity level for plant growth. The beneficial use of the end-products obtained in this study for crop production were attributed to postulations based on findings from this study and previous works on this subject matter, which include:

a.   stimulation of beneficial microorganisms in soil, which enhances soil fertility [25]

b.   possible increased photosynthetic rate in plants evidenced by increased photosynthetic pigments (chlorophylls a and b) [40]

c.   increase in soil buffering capacity [28]

d.   increased soil moisture retention capacity by reducing hydrophobicity tendency [29]

e.   positive soil temperature modifications that enhance soil nutrient bioavailability to plants [31, 40]

f.   formation of stable chelates with toxic metals such as Pb, Cu and Cd in order to reduce their bioavailability to plants [40]

g.   preferential exclusion of the chelated toxic metals from soil solution, allowing the plant nutrient elements to be assimilated into plant cells

h.   improvement of soil physicochemical properties via:

i.   increased aeration and water retention [29]

j.   activation of the macro and micro nutrients in soil in forms readily assimilated by plants [30, 40]

k.   improvement of plant root development and growth

l.   improvement of seed sprout of plants and subsequent shoot growth

m.   improved plant biomass production [26]

n.   enhanced soil nitrogen, phosphorus and potassium status for improved soil fertility

o.   acting as plant growth hormone, having positive stimulant action for plant growth [25, 26]

p.   improvement of soil permeability, promoting plant drought resistance [29]

q.   promotion of increased soil porosity and organic matter content, hence greatly promoting the microorganism activity and improving soil fertility.

Regarding leachate generation and management during the remediation exercise; fluid (leachate) produced as remediation progressed was recycled by incorporation into the biocell and used to regulate moisture content, thereby reducing water usage and conserving water resources. Expertise applied during the project ensured that at remediation project close-out, no isolated fluid system was actually produced. Nonetheless, the assessment of leachate effect on plant growth carried out in this work was to establish the fact that even in the event of accidental release of some fluid into the environment, there would be minimal risk to the receptor biotic community. More evaluations are still ongoing in this regard. Results from this study revealed that the leachate generated, though a concentrate, supported plant growth and when diluted with ordinary tap water gave a better support; reasons being that:

• toxic petroleum hydrocarbons in the contaminated drilling wastes have been destroyed to an acceptable level, evidenced by natural foamability of the concentrated leachate. Foamability would hardly occur if oil was still present

• leachate is also enriched with plant nutrients such as nitrogen, phosphorus and potassium

The process fluid, therefore, had some fertilizer value. The percentage decreases (1.50% and 23.45%) obtained for plant height and root length respectively, for the stock leachate was attributed to concentrated level of nutrients, confirmed by better performance of dilute leachate series. Naturally, in any formulated fertilizer, plant nutrients are applied at specified concentrations otherwise may hinder plant growth. Comparative evaluations of control system (soil treated with water only), stock leachate system (soil treated with leachate without any form of dilution) and systems with serial dilutions of the leachate (soil treated with leachate diluted with water by factors 1, 2, 3 and 4) revealed that the leachates were not toxic to receptor plants. The implication of this is that in the event of occasional spill of the leachate to the adjacent environment; dilution with water is, therefore, an adequate safety measure.

The ability of the end products to sustain the growth of green leafy vegetable: fluted pumpkin (*Telfairia ocidentis*) and root tuber crop, cassava (*Manihot esculenta Crantz*) and cereal crop (maize) is a demonstration of the utility of the remediation end product. It therefore stands that the use of CNB-Tech products as a biotechnological tool for hydrocarbon degradation in drilling waste converts these waste materials into non-toxic and potentially useful end products. In addition to the beneficial use of the remediation end-product for agricultural purposes, other possible utility options, shown in Figure 19, include:

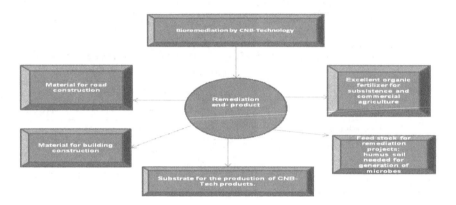

Figure 19. Potential utility of end - products from bioremediation using CNB-Tech products

- material for road construction

- material for building construction

- substrate for the production CNB-Tech bioremediation agents

- excellent organic fertilizer for subsistence and commercial agriculture

- feedstock for bioremediation projects

Table 6 is a comparative evaluation of economic, operational and environmental implications of thermal technologies as reported by [3] and CNB-Technology based on the results and learning from this study.

S/N	Thermal Technology	CNB-Tech
1.	Effective removal and recovery of hydrocarbons from solids	Effective removal of hydrocarbons from solid

S/N	Thermal Technology	CNB-Tech
2.	Possibility of recovering base fluid and end - product could be used for brick making	Effective recovery of free phase oil and end product has other uses apart from brick making
3.	Low potential for future liability	No future liability
4.	Requires short time	Time is relatively short
5.	High cost of handling environmental issues, since end-product dispersion would be below organic layer where vegetation growth is desired	Very minimized environmental issues
6.	Large volume of wastes is required to justify the cost of operation	Cost-effective for either small or large volume of wastes
7.	Requires tightly controlled process parameters	Does not require tightly controlled process parameters
8.	Heavy metals and salts are concentrated in processed solids	Reduces heavy metals and salts concentrations in process solid
9.	High operating temperatures can lead to safety risks	Low operating temperature. Operates at ambient temperature; modulation does not exceed 60°C.
10.	Requires several operators	Does not require several operators
11.	Process water contains some emulsified oils	Process water does not contain some emulsified oils
12.	Residue ash requires further treatment	No residue ash. End-product is clean soil
13.	End product is sterile and can no longer support plant Life	End product is fertile and can support microbial and plant Life

Table 6. Comparative evaluation between thermal technology and CNB-technology

5. Conclusions and recommendations

This study revealed that it is possible to harness natural, biodegradable and local resource materials of Nigeria origin; translate them to scientifically formulated products that can be utilized for efficient biodegradation of hydrocarbon polluted matrices such as oil-based mud and drill cuttings within a reasonable short period of 6 days. This technology thus converts hydrocarbon polluted oil-based mud and drill cuttings to beneficial end-products of high order reuse such as soil amendment, without the generation of secondary waste materials. Field-scale trial adopting CNB-Technology is recommended.

Acknowledgements

This project was carried out under full financial support of the Remediation Department, Shell Petroleum Development Company (SPDC), Port Harcourt, Nigeria through the University

Liaison Team of the company. The support of the Oil well Team of SPDC that facilitated the procurement of oil- based mud and drill cuttings is also acknowledged.

Author details

Iheoma M. Adekunle[1*], Augustine O. O. Igbuku[2], Oke Oguns[3] and Philip D. Shekwolo[2]

*Address all correspondence to: imkunle@yahoo.com

1 Environmental Remediation Research Group, Department of Chemical Sciences (Chemistry), Federal University Otuoke, Bayelsa State, Nigeria

2 Restoration of Ogoniland Project Team, Shell Petroleum Development Company, Port Harcourt, Nigeria

3 Remediation Team, Shell Petroleum Development Company, Port Harcourt, Nigeria

References

[1] United Nations Environmental Programme (UNEP), 2011. Environmental Assessment of Ogoniland. P.1-262. ISBN:978-92-807-9 Available on line at: http://postconflict.unep.ch/publications/OEA /UNEP_OEA.pdf

[2] Department of Health, Government of South Australia (DHGSA). Public Health Fact Sheet on Polycyclic Aromatic Hydrocarbons (PAHs): Health effects 2009 http://www.dh.sa.gov.au/pehs/PDF-files/ph-factsheet-PAHs-health.pdf

[3] Neff, M.M and Duxbury, MA. Composition, environmental fates, and biological effects of water based drilling muds and cuttings discharged to the marine environment: A Synthesis and Annotated Bibliography. Prepared for Petroleum Environmental Research Forum (PERF) and American Petroleum Institute. 2005. http://perf.org/pdf/APIPERFreport.pdf

[4] Gbadebo, A.M., Taiwo, A.M. and U. Eghele, U Environmental impacts of drilling mud and cutting wastes from the Igbokoda onshore oil wells, Southwestern Nigeria. Indian Journal of Science and Technology, 2010; 3(5), 504 -510.

[5] Environmental Protection Agency (EPA). An Assessment of the Environmental Implications of Oil and Gas Production: A Regional Case Study, 2008

[6] Osuji, L.C., Erondu, E.S and Ogali, R.E Upstream petroleum degradation of mangroves and intertidal shores: The Niger Delta Experience. Chemistry and Biodiversity, 2010: 7, 116 -128.

[7]   Knez, D., Jerzy, A, G and Czekaj Trends in the drilling waste management. Acts Montanistica Rocnlk, 2006:11, 80-83.

[8]   Morillon, A., Vidalie, J.F., hamzah, U.S., Suripno and Hadinota, E.K "Drilling and Waste management", SPE 73931, Intenationa; Conference on Health, Safety and Environment in oil and gas exploration and production, 2002: March 20-22

[9]   Zimmerman, P.K. and Rober, J.D Oil-based drill cuttings treated by landfarming. Oil and Gas J, 1991: 12, 81-84

[10] Rojas-Avelizapa, N.G., Roldan-carrillo, T., Zegarra-Martinez, H., Munez-Colunga, A.M and Fernandez-Linares A field trial for an ext-situ bioremediation of a drilling mud-polluted site. Chemosphere 2007: 66, 1595-1600.

[11] Frydda, S and Randle, J.B Case study: Biological treatement of Geothermal drilling cuttings. Proceedings World Geothermal Congress, Bali, Indonesia, 25-29, 2010: 1-3.

[12] Ouyang, W., Liu, H., Murygina, V., Yu, Y., Xiu, Z and Kalyuzhnyi, S Comparison of bio-augmentation and composting for remediation of oily sludge: A field-scale study in China. Process Biochemistry, 2005: 40, 3763 -3768.

[13] Vidali, M. Bioremediation: An overview. Pure and Applied Chemistry, 2001: 73(7), 1163-1173

[14] Jorgensen, K.S., Puutstinen, J and Suortt, A. –M Bioremediation of petroleum hydrocarbon-contaminated soil by composting in biopiles. Environmental Pollution, 2000: 107, 245-254.

[15] Department of Petroleum Resources. Environmental Guidelines and Standard for the Petroleum Industry in Nigeria, 2002

[16] Joel, O.F and Amajuoyi, C.A Determination of selected physicochemical parameters and heavy metals in a drilling cutting dump site at Ezeogwu–Owaza, Nigeria. J. Appl. Sci. Environ. Manage, 2009: 13(2), 27- 31.

[17] Okparanma, R.N., Ayotamuno, J. M Polycyclic aromatic hydrocarbons in Nigerian oil-based drill-cuttings; evidence of petrogenic and pyrogenic effects. World Applied Sciences Journal 2010; 11 (4): 394-400, ISSN 1818-4952.

[18] Nweke, C.O and Okpokwasili, G. C Drilling fluid base oil biodegradation potential of a soil Staphylococcus species. African Journal of Biotechnology 2003; 2 (9), pp. 293-295. http://www.academicjournals.org/AJB

[19] Ayotamuno, J.M., Okparanma, R, N and Araka, P.P Bioaugmentation and composting of oil-field drill-cuttings containing polycyclic aromatic hydrocarbons (PAHs). Journal of Food, Agriculture & Environment 2009; l.7 (2): 6 5 8 - 664. www.worldfood.net

[20] Okparanma, R.N Ayotamuno, J.M and Araka, P.P Bioremediation of hydrocarbon contaminated-oil field drill-cuttings with bacterial isolates. African Journal of Envi-

ronmental Science and Technology 2009 3 (5), pp. 131-140. Available online at http://
www.academicjournals.org/AJEST

[21] Ifeadi, C.N The treatment of drill cuttings using dispersion by chemical reaction
(DCR). A paper prepared for presentation at the DPR Health, Safety & Environment
(HSE) International Conference on Oil and Gas Industry in Port Harcourt, Nigeria.
2004.

[22] Adekunle, I.M., Ajijo, M.R., Omoniyi, I.T and Adeofun, C.O Response of four phytoplankton species in some sections of Nigeria coastal waters to crude oil in controlled
ecosystem. Int. J. Environ., Res., Iran, 2009; 4 (1): 65 -74 http://ijer.ut.ac.ir

[23] Adekunle, I.M and Onianwa, P.C Functional group characteristics of humic acid and
fulvic acid extracted from some agricultural wastes. Nigerian Journal of Science, Nigeria, 2001: 35 (1), 15 – 19.

[24] Adekunle, I.M Evaluating environmental impact from utilization of bulk composted
wastes of Nigerian origin using laboratory extraction test. Environmental Engineering and Management Journal 2010; 9 (5): 721 -729.: http://omicron.ch.tuiasi.ro/EEMJ/

[25] Adekunle I.M., Adekunle, A.A., Akintokun, A.K., Akintokun, P and Arowolo,T.A
Recycling of organic wastes through composting for land applications: a Nigerian experience. Waste Management & Research 2011; 29 (6): 582 – 593. DOI: 10.1177/ http://
wmr.sagepub.com/content/29/6/582.abstract

[26] Adekunle, I.M Bioremediation of soils contaminated with Nigerian petroleum products using composted municipal wastes. Bioremediation Journal, 2011; 15 (4):
230-241,   DOI: 10.1080/10889868.2011.624137.   http://dx.doi.org/
10.1080/10889868.2011.624137

[27] Adekunle I.M., Oguns, O., Shekwolo, P.D., Igbuku, O.O and Ogunkoya, O.O Assessment of population perception impact on value-added solid waste disposal in developing countries, a case study of Port Harcourt City, Nigeria. In: Xiao-Ying, Y (Ed)
Municipal and Industrial Waste Disposal. Intech; 2012, p177-206.

[28] Adekunle A. A., Adekunle, I.M., Igba, T. O Assessing the effect of bioremediation
agent from local resource materials in Nigeria on soil pH. Journal of Emerging
Trends in Engineering and Applied Sciences 2012; 3 (3) 526-532. http://jeteas.scholar-
linkresearch.org/articles/Assessing%20the%20Effect%20of%20Bioremediation
%20Agent.pdf

[29] Adekunle A.A., I.M. Adekunle and Igba, T.O Impact of bioremediation formulation
from Nigeria local resource materials on moisture contents for soils contaminated
with petroleum products. International Journal of Engineering Research and Development 2012; 2(4) 40-45 http://www.ijerd.com/paper/vol2-issue4/F02044045.pdf

[30] Adekunle A.A, Adekunle, I.M. and Igba, T.O Assessing and forecasting the impact of
bioremediation product derived from Nigeria local raw materials on electrical conductivity of soils contaminated with petroleum products. Journal of Applied technol-

ogy in Environmental Sanitation 2012; 2 (1) 57 -66. http://www.trisanita.org/jates/atespaper2012/ates09v2n1y2012.pdf

[31] Adekunle A.A., I. M. Adekunle and Igba T. O Soil temperature dynamics during bio-remediation of petroleum products using remediation agent for Nigerian local re-source materials. International Journal of Engineering Science and Technology 2012; 1 (4): 1-8. http://www.ijert.org/browse/june-2012-edition

[32] Association of Official Analytical Chemists (AOAC), Official Method and Analysis of The Association oh The Official Analytical Chemists 11th Edition Washington D C, 1970.

[33] Finar, I.L Organic Chemistry, volume I The Fundamental principles. 6th Ed, Long-man, 1973.

[34] Liu, W., Luo, Y and Teng, Y Bioremediation of oily sludge-contaminated soil by stim-ulating indigenous microbes. Environ Geochem health 2010: 32, 23 -29.

[35] Ayotamuno, J.M., Okparanma, R.N., Davis, DD and allagoa, M. PAH removal from Nigerian oil-based drill-cuttings with spent oyster mushroom (Pleurotus ostretus) substrate. Journal of Food, Agriculture and Environment 2010: 8 (3 &4), 914 -919.

[36] Rojas-Avelizapa, N.G., Roldan-Carrillo, T., Zegarra-Martinez, H., Munoz-Colunga, A.M and Fernadez-Linares A field trail for an ex-situ bioremediation of a drilling mud-polluted site. Chemospher, 2007: 66, 1595 – 1600.

[37] Martin, J.A., Moreno, J.L., Hernandez, T and Garcia, C Bioremediation by compost-ing of heavy oil refinery sludge in semiarid conditions. Biodegradation, 17:, 251 – 261.

[38] Al-Nasrawi, H Biodegradation of Crude Oil by Fungi Isolated from Gulf of Mexico. J Bioremed Biodegrad 2012; 3:4

[39] Mandal, A.K., Sarma, P. M., Singh, B., Jeyaseelan, C.P., Channasshettar, V.A., Lal, B and Datta, J bioremediation : an environment friendly sustainable biotechnological solution for remediation of petroleum hydrocarbon contaminated waste. ARPN Jour-nal of Science and Technology, 2012: 2, 1-12

[40] Stevenson, F.J Humus Chemistry, 2004. Wiley & Sons

[41] Obayori, O.S., Ilori, M.O., Adebusoye, S.A., Amund, O.O and Oyetibo, G.O Microbial population changes in tropical agricultural soil experimentally contaminated with crude petroleum. African Journal of Biotechnology, 2008: 7 (24), 4512-4520.

[42] Corwin, D.L and Lesch, S.M. Apparent soil electrical conductivity measurements in agriculture. Computers and Electronics in Agriculture, 2005: 46, 11–4

Aerobic Biodegradation Coupled to Preliminary Ozonation for the Treatment of Model and Real Residual Water

P. Guerra, J. Amacosta, T. Poznyak, S. Siles,
A. García and I. Chairez

Additional information is available at the end of the chapter

1. Introduction

1.1. Sequence of water treatment methods

Residual and waste water have become a problem of paramount importance in modern societies [1]. Recently, the number of proposals to solve this issue has incremented importantly [2]. Several methods were proposed since thirty years ago using a wide variety of physical, biological and chemical principles. Biological treatments are cheap and environmentally friendly [3]. Nevertheless, they require a long time to eliminate pollutants and they are limited by the toxicity and initial concentration of the water sample that must be treated [4]. On the other hand, chemical treatments are capable to promote the faster decomposition for a wide range of toxic compounds [5]. Despite this adequate performance to decompose organics dissolved in water, they are hundreds or thousands of times more expensive than pure biological methods [6-11].

1.2. Ozonation followed by biodegradation

Just some years ago, the attractive features of both methods (biological and chemical) have attracted attention to develop more advanced schemes to manage more toxic and complex pollutant mixtures [12-13]. Indeed, remarkable results have come from a sequence of treatments (usually called trains of treatments) using the combination of several individual options. Regularly, the treatment trains are using a sequence defined by a physical method followed by the biological scheme and finally, one last chemical process completes the treatment.

However, this arrangement does not always work efficiently when the initial pollutant mixture has complex composition or they are very toxic. The stage that is usually compromised by this aspect is the biological one [14-15].

Several mixed processes have been recently proposed including a process based on chemical oxidative compounds plus biological based decomposition course [12-18]. Among others, oxygen injected with high pressure, ozonation [13], catalytic and photocatalytic processes and others have been tested to perform the chemical decomposition [16-18]. Most of these treatment methods have important advantages but also have important drawbacks. These problems can be classified into two main areas: the first one contains all economic aspects associated with high cost required to implement these treatments, the second one includes all troubles associated to the resources needed to complete the transformation from very toxic compounds to simpler ones that can be considered as no toxicity and no hazardous [19]. Nevertheless, these drawbacks may be solved by biologically based treatments.

Nowadays, a different way of thinking has emerged to improve the efficiency of waste and residual water: changing the order of treatments to include a chemical pre-treatment before the biological process. The idea is to reduce the complexity as well as the toxicity of the organic mixture of chemical methods. Theoretically, this condition must have a positive effect on the microorganism's efficiency to decompose the simpler and less toxic organics.

Beltran et al. [13] reported that combined ozonation and aerobic treatment increased the removal efficiency from 82% or 76% for the COD or the total phenolic content, respectively. Benitez et al. [20] demonstrated the COD removal for wine vinasses containing organic matter and aromatic compounds was enhanced (from 27.7% to 39.3%), when the combined ozonation and biological process was used. Aparicio et al. [21] reported the use of combined wastewater treatment set up in a resin-producing factory. After biological treatment of the ozonated effluent, the organic carbon and nitrogen removal was increased from 27 to 97% and from 27 to 80%, respectively.

The possible benefits coming from the combination of pre-treatment with ozone and a sequential biodegradation are almost evident; however, there are still several questions about this procedure. For example, what time is adequate to move the organics mixture from the ozonation reactor to the biological one or what conditions should be set-up for both reactions still remain as open problems. Another important issue that must be explained is what conditions must fulfill the microorganism strains to handle the pollutant mixture produced by the preliminary ozonation. This is an important aspect conditioned to the composition of the mixture supplied to microorganisms that can modify the organics elimination by biodegradation. Moreover, there is just a few of works describing what type of microorganisms is responsible for the elimination of residual compounds after ozonation [22]. In recent reports, [23-29] catechol, hydroquinone and several low weight organic acids have been recognized like the main byproducts obtained after ozonation of phenol and its chlorinated derivatives. Nevertheless, what relative concentration of each byproduct is the most adequate to construct the combined process including ozonation and biological reaction has not been determined yet.

1.3. Phenol and its chlorinated derivatives as artificial wastewater

Phenol and its chlorinated derivatives are simple examples of how the biodegradation can be efficient or not for closely related pollutants [26, 27]. It has been broadly reported the efficient decomposition of phenol by biological strands. Many methods of eliminating phenol and its derivatives using chemical and biological systems have been studied [23-24]. Biological systems are environmentally friendly, low-cost technologies that can be successfully used to remove phenols by using different microbial strains but with pure cultures.

However, when any chlorophenol is exposed to the same microorganisms, the toxicity of this compound reduces the decomposition efficiency to 20 or 30 % compared to the same conditions observed in phenol treatment.

On the other hand, most of advanced oxidation processes can remove the chlorine atom from the chlorophenols in few minutes or even seconds [5]. Therefore, the possible sequential treatment based on ozone followed by biodegradation can use the advantages offered by this couple of water treatment schemes. Indeed, phenol's ozonation generates simpler organic acids that can be assimilated by microorganisms [14-15].

1.4. Lignin and its derivatives as real wastewater

A more complex situation arises when the pollutants in residual water are toxic and also with complex structure. As an example, pulp and paper industry wastewater mainly contains lignin and its derivatives (chlorinated phenolic compounds, resin acids, dioxins and dioxin-like compounds and many others) [29-32].

Lignin is a three-dimensional biopolymer which confers the resistance to stress, protects the plant from the microbial enzymatic hydrolysis and also acts as a binder of the fibers of cellulose and hemicellulose in the wood [33,34]. Is formed by the coupling of three monolignols (paracoumaryl alcohol, coniferyl alcohol and sinapyl alcohol) and with some functional groups (ROH, ϕOH, ROMe, RR'C=O, RCOOH, RSO_3R', etc.). The extractable products like fatty acids, phenols, terpenes, steroids, waxes, tannins and resinic acids, also found in the wastewaters, confer the physicochemical properties of each plant such as color, smell, strength, hardness [37-39].

Lignin and its derivatives have shown to be a very complex, toxic mixture with mutagenic and teratogenicity activities. Biological treatment of wastewater with these residuals has partial efficiency [39-41].

1.5. Motivation and contribution of this study

In this chapter, a combined method to treat residual water is proposed. The treatment is based on the preliminary action of ozone followed by the biological treatment using a microorganism consortium. Two water samples were used to evaluate the combined treatment: a model mixture of chlorophenols and residual water obtained from the paper industry after the bleaching step from the Kraft process. This selection was done to illustrate the efficiency of the combined process using ozone and biodegradation in a row.

2. Materials and methods

2.1. Model and real residual water

Model solutions were artificially prepared with 4-Chlorophenol (4-CPh) or 2,4-Dichlorophenol (2,4-DCPh) (120 mg/L) as model solutions. All these chemical products have 99% purity.

For the real residual water, the sample was obtained from a Kraft process in the bleaching step; collected at 4 °C and sterilized in autoclave under the temperature of 121 °C and a pressure of 15 pounds. The mixture was characterized by simple analytical methods based on UV/VIS spectroscopy.

These two polluted water samples were treated by ozonation, aerobic biodegradation and the combination of both processes (ozonation followed by biodegradation). The biodegradation was developed using a microbial consortium acclimated to the particular composition of carbon source remaining in the reactor after/before ozonation [42-44].

2.2. Ozonation procedure

The ozonation treatment was carried out in a semi-batch glass reactor (250 mL). Ozone concentration at the reactor's input was 30 mg/L. The maximum ozonation time was 60 min for both the real and artificial wastewater. The ozone/oxygen mixture was injected through a ceramic porous at the inferior part of the reactor with a flow of 0.5 L/min. The ozone was produced by the ozone generator HTU500G 'G' (corona discharge type, "AZCO" INDUSTRIES LIMITED, Canada). The Ozone Analyzer BMT 963 "S" (BMT Messtechnik, Berlin) provides the ozone detection in the gas phase at the reactor output. This information was used to perform the ozone monitoring, to control the ozonation degree and to study the ozone decomposition. The ozone concentration was sampled by a data acquisition system implemented in a regular computer (Figure 1).

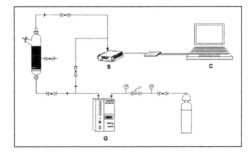

Figure 1. Simple scheme of the ozonation setup including the reactor where the ozonation is carried out. The ozone concentration produced in (G) is monitored in the UV sensor (S). A data acquisition board is connected to a personal computer to register the ozone concentration.

The ozonation of was carried out using two different pHs: 7 (for model and real water) and 12 (only model solution). Ozonation of residual water samples was carried out for diluted solutions (1:10). This change promotes the reduction of reaction time and helped to decrease foaming [16].

2.3. Microbial culture, mineral media

For the biological treatment, different microbial consortia were cultivated and acclimated during 6 months (by a fill-and-draw procedure) to the specific carbon source (model chlorophenols and real water solutions with and without previous ozonation).

The mineral media used for all the experiments contains (g/L): 3.0 $(NH_4)_2SO4$, 0.6 KH_2PO_4, 2.4 K_2HPO_4, 1.5 $MgSO_4 \bullet 7H2O$, 0.15 $CaSO_4$, and 0.03 $FeSO_4$. A mixed microbial culture from a biofilter used to remove aromatic compounds and gasoline vapors (Dr Revah's Laboratory, Universidad Autonoma Metropolitana Iztapalapa, Mexico) was independently adapted for three months to phenol (100 mg/L) and to a mixture of oxalic and formic acids (100 mg/L each) in mineral media.

The mixed culture was cultivated in an Erlenmeyer flask of 1 L with 500 mL of mineral media. These compositions were inoculated with 50 mL of the microbial mixture. Reactors were kept at ambient temperature and shaken in an orbital shaker at 200 RPM. The mineral media was also kept invariable for these experiments. In all the studied samples, the biomass amount and the organic degradation as well were periodically measured in triplicate.

The cultures were harvested between 24 and 30 h, corresponding to the exponential growth phase and then used for the model and real water treatment (phenol and chlorophenols mainly) and their corresponding ozonation products (catechol, hydroquinone and organic acids mainly) [46].

For the biological treatment, different microbial consortia (set of several microorganisms with different species) were cultivated and acclimated during 6 months (by a fill-and-draw procedure) to the specific carbon source (model chlorophenols and real water solutions with or without previous ozonation). In this study, the biological media is composed of a complex consortium of the microbial population previously identified [22] by the extraction of DNA samples using an Easy-DNATM Kit (Invitrogen, USA) [44].

Inoculums of the corresponding microbial consortium were added into the batch reactor containing the model solution or the real water (with or without previous ozonation). Reactors were kept at ambient temperature and shaken in an orbital shaker at 200 RPM. Chlorophenolsenols (from model solution) and real water's components concentration, as well as ozonation products concentration in reactors and the biomass amount were periodically measured by triplicate.

2.4. Analytical methods

Several analytical methods were used in order to characterize, identify and quantify the samples. UV Spectroscopy (Lambda 2S, Perkin Elmer) was used for monitoring the global behavior of ozonation (λ=260 and 210 nm) and biodegradation (λ=210 nm) of real water as well as the microbial growth (OD600).

The control of chlorophenols or the components of real water decomposition, as well as the intermediates and final products formed in the ozonation step was made by high performance liquid chromatography (HPLC), (Series 200, Perkin Elmer) equipped with UV-VIS detector series 200. Two wavelengths were periodically monitored (210 nm and 270 nm). Analytical details are shown in Table 1.

Analysis conditions	Compounds	
	Phenols	Organic acids and real water
Column	Platinum C-18 (Alltech), 250 x 4.6mm	Prevail Organic Acid (Grace), 150 x 4.6mm
Mobile Phase	60:40 (water : methanol)	KH_2PO_4 25 mMol in water (pH = 2.5)
λ(nm)	210	
Flow rate	1mL / min	
Sample volume	30µL	

Table 1. HPLC analysis conditions

The study was made on raw material and samples of study, both in the stage of ozonation of biodegradation. Identification and qualitative determinations were made taking into account the retention times of components and the quantitative analysis by integration of signals, in relation to the corresponding calibration curve.

3. Results

For both kind of waters, model solution or real water, three processes were evaluated: ozonation, biodegradation (without ozonation) and the combined treatment (ozonation followed by biodegradation). Those are described below.

3.1. Ozonation

3.1.1. Model solution preliminary ozonation

Chlorophenols (CPhs) decomposition was faster at pH 12 (8 and 5 minutes) than pH 7 (15 and 8 minutes) for 4-CPh and 2,4-DCPh, respectively. Some by-products like catechol, hydroquinone, oxalic and formic acids were formed. All these are some of the products identified in CPhs ozonation [5]. Besides, some other ones were observed but they could not be identified, however, they were monitored by HPLC and referred as non- identified phenolic compounds and organic acids. During ozonation, both identified and non-identified phenolic compounds were rapidly decomposed, while oxalic and formic acids were mostly accumulated during the whole reaction period. The maximum concentration detected for the different ozonation conditions were previously published [5]. All these compounds constituted the carbon source for adapted bioprocess applied at the next step. Then, the percentage of CPhs decomposition and the by-products accumulation/decomposition was considered to stop the ozonation and carry out the biodegradation step.

3.1.2. Real residual water ozonation

In the case of real residual wastewater ozonation, a significant decrease of organic compounds concentration was followed in the UV spectra (Figure 2). Lignin derivatives was followed in the UV region of λ=260 nm and organic acids in the region of λ= 210 nm.

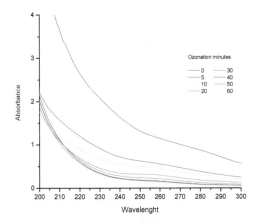

Figure 2. Variation of UV-VIS spectrum in the ozonation process.

A significant decrease in the region of lignin derivatives (94%) and in the corresponding organic acids (83%), which also tends to decrease in absorbance is depicted. At the end of ozonation no longer variation in the UV spectra is observed. Therefore, the susceptible organic matter to be ozonated has been completely reacted with ozone.

The identification of organic compounds in the original sample and also during the ozonation was done by the HPLC technique. The decomposition dynamics during the reaction with ozone was determined. Moreover quantification of accumulated products is gotten. Main byproducts were hydroquinone, catechol and simple organic acids such as maleic acid and several unidentified compounds.

The main recalcitrant accumulated product during ozonation was oxalic acid. Indeed, its presence was observed from the beginning to the end of ozonation. Its relative importance for the sequent biodegradation motivated its quantification by HPLC (Figure 3).

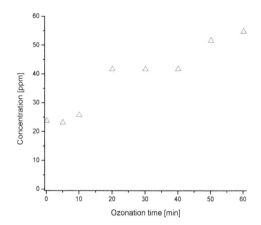

Figure 3. Quantification of oxalic acid during the ozonation of real wastewater.

As it can be seen, oxalic acid was contained in the real residual wastewater before any treatment (23 mg/L), but after the first 30 minutes 43 mg/L of this acid was detected. After 60 minutes of ozonation the acid concentration was increased up to 55 mg/L.

3.2. Biodegradation

3.2.1. Biological treatment of model solution (without ozonation)

Poor degradation of CPhs was observed in the case of non ozonated samples (30% and 40% for 4-CPh and 2,4-DCPh, respectively) after 10 days. No matter the microorganisms were previously acclimated to CPhs, the initial concentration was toxic enough to inhibit biodegradation, which is one of the biological treatments principal disadvantages. The biodegradation of ozonation products identified during ozonation was also tested. The minority compounds were eliminated obeying the following order of elimination: phenol>catechol>hydroquinone. For final products, a mixture of oxalic and formic acids with a concentration of 100 mg/L of each one was also tested. Microorganisms were able to eliminate both compounds during two days. These results are very important because they demonstrated that: highly toxic substrates, which cannot be eliminated by bioprocess, but they can be easily degraded by ozone and transform them into several compounds.

These results demonstrated that highly toxic substrates which cannot be eliminated by bioprocess are in the model solution, but they can be easily degraded by ozone and transform them into several compounds. Additionally, an acclimated consortium whit capability to eliminate the ozonation products was produced. So, it was expected that ozonation products were easier to be eliminated by biodegradation than original CPhs. Therefore, one can expect that combined treatment is more efficient than individual ones.

3.2.2. Biological treatment of real residual water (without ozonation)

The real residual water biodegradation showed a similar behavior to model solution one. It showed a partial decomposition of the organic compounds up to 20 % after five days due to the complexity and heterogeneity of this sample without ozonation. Therefore, organic matter degradation is poor (Figure 4).

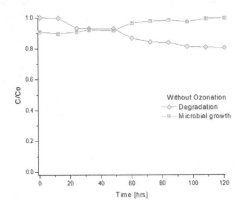

Figure 4. Growth dynamics of biomass and decomposition of organic matter without pre-treatment with ozone.

Nevertheless, microbial growth was obtained but probably because of the consumption of oxalic acid present in non ozonated sample. Taking into account that oxalic acid (which could be eliminated by microorganisms) is in real water (before ozonation) and this compound is formed further in ozonation, it is expected that combined treatment was successful. In general, by comparing in Figure 4 the growth / degradation dynamics appears in normalized form to compare them. Both, the increase of the Optical Density measured at 600 nm (corresponding to microorganisms grow), and the decrease for the consumption of the source of carbon are shown together.

First 20 minutes are characterized by a partial decomposition of the organic compounds up to 10 %. However, after 50 minutes, an additional decomposition around 10% is obtained but no more pollutants elimination was achieved until the end of the experiment (120 hours).

As it was expected, the organic matter decomposition is not significant. Only 20% of the total organic matter was eliminated due to the complexity and heterogeneity of the wastewater sample. As expected, microbial growth was not considerable because the pollutants nature as well as their concentration in the sample.

3.3. Combined treatment

3.3.1. Biological treatment of model solution after ozonation step

Taking into account the results obtained in previous sections, several ozonation conditions were chosen to test combined treatment. Three principal aspects were considered to choose pre ozonation conditions: the reduction (or complete elimination) of CPhs concentration, the accumulation of phenolic compounds and the final production of organic acids. As phenolic compounds are formed since the beginning but rapidly decomposed during ozonation, its presence was evaluated during biodegradation, in order to consider the needy to continue ozonation till remove them. Table 2 shows these pre ozonation conditions with corresponding ozonation products concentration.

Identified Compounds	Concentration (mg/L)						
	Ozonation time (min)						
	4-CPh				2,4-DCPh		
	pH 7			pH 12	pH 7		pH 12
	10	15	30	5	5	8	5
4-CPh	9	-	-	18	-	-	-
2,4-DCPh	-	-	-	-	18	-	-
Oxalic Acid	10	15	27	30	6	9	48
Formic Acid	154	137	43	39	36	50	54

Table 2. Ozonation conditions for the sequential treatment

In order to measure the biodegradation of pre-ozonated samples, the total area registered in HPLC chromatograms (under each condition reported in Table 2) was integrated. A normalized version of this area was used to evaluate the evolution of pollutants decomposition (before biodegradation). For the model sample, total degradation of two kinds of substrates (phenolic compounds and organic acids) was monitored. We could corroborate that the amount of each substrate (phenolics or organic acids) accumulated in ozonation has a very important influence during biodegradation step.

Figures 5, 6, 7 and 8 show the biodegradation on combined treatment of organic acids and phenolic compounds. For both kinds of substrates, during the first two days of biodegradation, it is remarkable the influence of ozonation conditions (Table 2). Indeed, it is a direct relationship between the formation-decomposition dynamics of by-products ozonation and the biodegradation facility of different kind of substrates.

Remembering the separated processes describe above, during ozonation, the phenolic compounds tend to accumulate during the first stage of treatment and after that, they tend to decomposed (Figures 5 and 6 for 4-CPh and 2,4-DCPh respectively). On the other hand, organic acids are formed since the beginning and they tend to accumulate (being the oxalic acid the most representative acid at the end of ozonation). On the other hand during biodegradation the oxalic and formic acids demonstrate to be easier to eliminate than phenolic compounds (Figures 7 and 8 for 4-CPh and 2,4-DCPh respectively). Under this perspective, it is very clear why during the first days of biodegradation in combined treatment; the phenolic compounds are more degraded if the ozonation time was higher (under pH 7). Without a doubt, ozonation partially degraded these compounds, during biodegradation and then, they are less complicated to be eliminated by microorganisms. This last idea is confirmed by the decomposition dynamics observed for biodegradation of 4-CPh and 2,4-DCPh ozonated under pH 12. Similar behavior was observed for organic acids biodegradation.

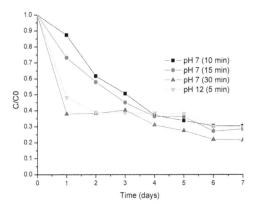

Figure 5. Biodegradation of phenolic compounds accumulated during 4-CPh ozonation

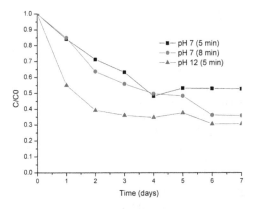

Figure 6. Biodegradation of phenolic compounds accumulated during 2,4-DCPh ozonation

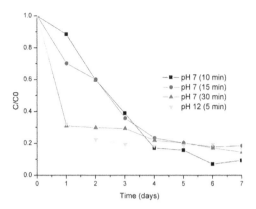

Figure 7. Biodegradation of organic acids accumulated during 4-CPh ozonation

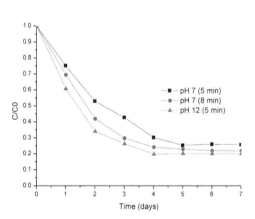

Figure 8. Biodegradation of organic acids accumulated during 2,4-DCPh ozonation

Biodegradation profiles are similar for phenolic compounds and organic acids. These organic acids are easier to eliminate than phenolic compounds. The 4-CPh decomposition degree for each substrate was between 81-90% and 70-78% for organic acids and phenolic compounds, respectively. On the other hand, 2,4-DCPh have decomposition degree between 74-80% and 47-69% for organic acids and phenolic compounds correspondingly. Table 3 shows a summary of elimination efficiency obtained for each pre ozonation condition.

It is important to notice that no matter the ozonation conditions, resulting by-products (phenolic compounds as well as organic acids) were metabolized by the microbial population since the beginning of the biotreatment, some of them faster than the others, but thank to

Compound	pH	Ozonation time (min)	% Biodegradation	
			Organic Acids	Phenolic Compounds
4-CPh	7	10	90	70
		15	81	71
		30	85	78
	12	5	86	70
2,4-DCPh	7	5	74	47
		8	78	64
	12	5	80	69

Table 3. Percentage biodegradation of 4-CPh and 2,4-DCPh pre-ozonated.

previous acclimation the microbial consortia is able to consume ozonated substrates. It is remarkable that residual 4-CPh (not fully eliminated during ozonation) remaining in ozonated samples during 10 and 5 minutes under pH 7 and 12, respectively was eliminated in biodegradation during the first day of treatment.

Analyzing the individual biodegradation profile of each compound formed during previous ozonation, a serial degradation is inferred. This means than some compounds were preferably consumed (in the earlier days of biodegradation) because of their energetic wealth or ease of degradation. When those organics were depleted, microbial consortium was able to metabolize all others (data not shown).

As it was previously mentioned, a microbial consortium was acclimated to specific carbon source, in this case, ozonation by-products. So, microbial population had developed specific abilities to degrade ozonation products.

Figure 9 shows the global biodegradation behavior in combined treatment for 4-CPh ozonated 10 minutes at pH 7. Substrates elimination and microbial growth are parallel. They were presented in a normalized way as diminution of the HPLC area (phenolic compounds and organic acids) and optical density increase.

The correlation between the optical density measured at 600 nm and the pollutants decomposition suggests organic matter integration in the biomass concentration. Between the day 0 and 1, poor degradation is obtained and in agreement, poor microbial growth. Between day 1 and 4, the major organic acids depletion and an important one for phenolic compounds is observed, so a second growth step appears as a result of metabolism of these substrates. Between day 4 and 7 the third growth step is observed as a result of metabolism of residual substrates (phenolic compounds and organic acids consumed until the end). It is remarkable that biodegradation trends of both substrates are simultaneous. However, from the second day, consortium shows an evident preference for organic acids (90% removal) over the phenolic compounds (70% elimination). Those dynamics were similar for all the combined treatments considered in this study (data not shown).

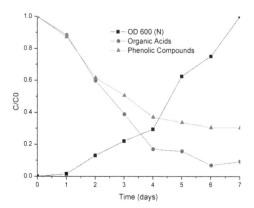

Figure 9. Substrates degradation and microbial growth during biodegradation of pre-ozonated 4-CPh (pH 7, 10 min).

Microbial growth in ozonated substrates was followed by optical density at 600 nm. This analysis was done when different pre-ozonation times (10, 15 and 30 minutes for pH is 7.0 and 5 minutes for pH is fixed to 12.0) were considered.

When 4-CPh is ozonated, microbial growth was faster when pH is 7.0 and ozonation time is fixed to 30 minutes. This accelerated biomass accumulation is associated to the major pollutants decomposition. Moreover, when pH was fixed to 12 and the reaction time was 5 minutes, the lower biomass velocity growing was achieved. If pH was fixed to 7.0, when reaction time was 15 minutes, the lag phase was delayed more than any others (Figure 10). This is explained by the accumulation of phenolic compounds under this reaction conditions. This is confirmed by the faster biomass accumulation when ozonation time was 10 minutes and phenolic compounds were not so higher than the previous case.

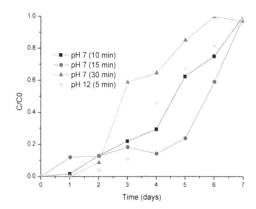

Figure 10. Biomass growth in biodegradation of organic acids accumulated during 4-CPh ozonation

When 2,4-DCPh was ozonated, the faster biomass accumulation was obtained when pH was 7.0 and the ozonation time was 8.0 minutes. Once again this condition coincides to the case when lower phenolic derivatives were observed in the reactor. Indeed, when pH was 7.0 with reaction time was 5 minutes and when pH was 12 and reaction time was 5, a lower biomass accumulation was determined. This is in agreement to the previous case, because under these two cases higher phenolic concentrations were observed. As one can understand, when phenolic compounds were at these levels, no important organic acids were in the reactor. This is a contrary situation to the case when phenolics compounds were at their minimum (among other studied cases) concentrations.

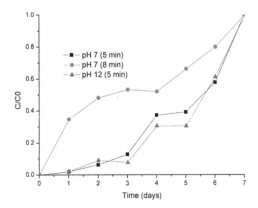

Figure 11. Biomass growth in biodegradation of phenolic compounds accumulated during 2,4-DCPh ozonation

Finally, Figures 12 and 13 show the UV spectra obtained after the combined treatment. These treatments were developed according to conditions presented in table 2.

Major elimination by combined treatment was obtained when most of the not identified compounds were decomposed, respectively (4-CPh ozonated during 30 min at pH7). This result is in agreement to the higher biomass accumulation observed in Figure 10.

Pre-ozonation conditions had an important influence on the overall degradation. When most of the no identified compounds were decomposed during ozonation (in the subsequent biodegradation step) the UV spectra is very close to the control case (mineral media, absence of contaminant).

It can be noticed that ozonation by-products were not as toxic as the original ones. This is explained because they were simultaneous consumed and corresponding to the microbial growth. In all the studied combined treatments, organic acids were the preferred substrates,

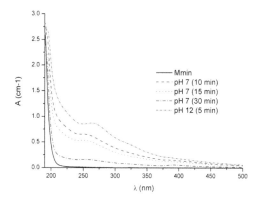

Figure 12. Effect of combined treatment in UV spectra of 4-CPh.

as they were assimilated faster than phenolic compounds. Indeed, phenolic derivatives have shown to serve as inhibitors of the biomass growing.

When 2,4-DCPh was ozonated, a similar condition to the previous one is recognized. The correlation between the biomass growing (showed in the Figure 11) and the phenolics concentrations was confirmed. Moreover, if pH was fixed to 7 and ozonation time was 8 minutes, an important organic matter decrease was observed (Figure 13).

A slighty difference between this case and the previous one shoukd be remarked: the higher organic matter decomposition is gotten. This is explained by the reaction mechanism that has been identified in preliminar studies. In this case, toxicity of byproducts can have a remarkable role on the biomass growing.

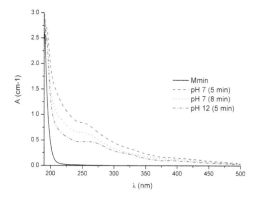

Figure 13. Effect of combined treatment in UV spectra 2,4-DCPh.

3.3.2. Biological treatment of real residual water after ozonation step

Two important aspects were observed to improve biodegradation in combined treatment of model solution: 1) phenolic compounds must be eliminated to the lowest achievable value and 2) short chain organic acids should be accumulated as much as possible. Considering these facts and the real residual water ozonation study, two ozonation conditions were chosen to test the suggested combined treatment: 30 and 60 minutes under pH 7. These conditions were selected from the study regarding model sample.

Figures 14 and 15 showed the global biodegradation behavior in combined treatment for real water ozonated during 30 and 60 minutes. In a similar fashion to the model solution, substrate elimination and microbial growth are parallel. To compare the results obtained for the model sample and the real wastewater, biomass growing and substrate are presented in normalized way. They are presented as diminution of UV spectra signal and optical density increases correspondingly.

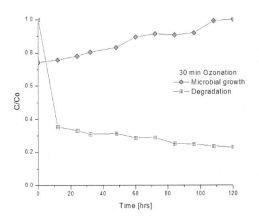

Figure 14. Growth dynamics of biomass and decomposition of organic matter after 30 minutes of ozonation.

During the first 12 hours of biodegradation for real water previously ozonated 30 minutes, more than 60% of organic matter has been metabolized by microorganisms. Moreover, after 5 days, 82% of the initial substrate was removed. In the same way, to the real water, a second sample was ozonated 60 minutes. During the first 12 hours, 78% of the substrate was eliminated and after 5 days, 83% of substrate was metabolized. As it can be observed, the degradation of all compounds that started from the first minutes.

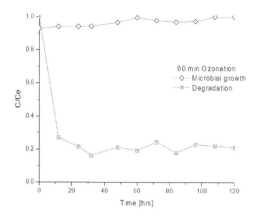

Figure 15. Growth dynamics of biomass and decomposition of organic matter after 60 minutes of ozonation.

This behavior is explained by transformation of pollutants performed by microorganisms. In particular, less toxic substrate as short organic acids were again preferred by them. A decomposition degree of 72%, after 120 hours was gotten.

In the same way for the sample ozonated by 60 minutes, after the first 12 hours, initial substrate was eliminated 78%. When the biodegradation was stopped, an organic matter decrease of 83% was obtained. On the other hand, it is necessary to pay special attention to the almost unchangeable behavior after 30 hours. This is attributed to the formation of some specific products of biodegradation, which can perhaps be subsequently consumed by microorganisms (tendency to continue decreasing). Remembering that most of the ozonated real water is composed of oxalic acid and similar short chain acids. Therefore, the substrate consumption in real water samples is having a similar portrait to organic acids consumption in model solution, it means, since the biodegradation begins, all these acids are metabolized.

Figure 16 shows the dynamics of the three systems after 5 days of biodegradation. This time was selected as the maximum time for the bioprocess. This study presented a removal of organics of 85% for sample ozonated by 30 minutes and 89% for the sample ozonated by 60 minutes.

Finally, in a quantitative way the global effect of biodegradation is the elimination of oxalic acid that was previously formed during ozonation. It is clearly observed that there is a decrease in the concentration of oxalic acid during biodegradation.

In the same way, 30 minutes of ozonation as pre-treatment is more efficient than the other one. They both have similar effects but the suggested one is using half of time for treatment with ozone. As result, final costs of combined treatment are reduced.

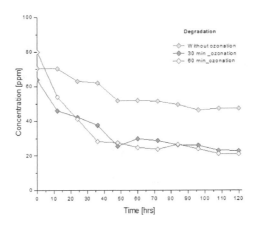

Figure 16. Decomposition dynamics of the three systems (without ozonation, 30 and 60 min of ozonation) after 5 days of biodegradation.

4. Conclusions

The combine residual water treatment using ozone before biodegradation seems to be an interesting option to eliminate more complex and toxic contaminants. The combined treatment may handle the aforementioned type of organics mixtures but with less cost than the pure chemical method and with a shorter treatment period than the biological procedure. For the model residual water, the preliminary ozonation decompose organics in the complex mixture and produce more biodegradable species like organic acids. Longer ozonation times are better if one takes into account the decomposition and accumulation dynamics of both, phenolic compounds and organic acids. When most of the no identified compounds were decomposed during ozonation, after the biodegradation step, the UV spectra was very close to the mineral media one in absence of contaminant. Then a major grade of mineralization after combined treatment is obtained. The previous acclimation of the consortium also showed an improvement of the complete treatment scheme for model residual water. The mineralization of ozonation by-products was confirmed by the microbial growth (in aerobic biodegradation, microbial growth is always accompanied by CO_2 production). In the real residual water, the obtained results confirmed the completely decomposition of toxic residues after 60 minutes of ozonation. The decomposition dynamics of lignin derivatives and chlorinated phenols are proportional to the formation of the oxalic acid. This is the main final product of ozonation. Degradation dynamics of these compounds are shown as well as the formation of the oxalic acid.

Acknowledgements

This work was supported through funding provided by CONACYT grants 49367, 60976. Also, authors thanks to the IPN project SIP-20120406.

Author details

P. Guerra[1], J. Amacosta[1], T. Poznyak[1], S. Siles[1], A. García[2] and I. Chairez[3*]

*Address all correspondence to: isaac_chairez@yahoo.com

1 Superior School of Chemical Engineering, National Polytechnic Institute (ESIQIE-IPN), Mexico

2 Instituto Tecnológico de Estudios Superiores de Monterrey, Campus Guadalajara (ITESM), Mexico

3 Professional Interdisciplinary Unit of Biotechnology, National Polytechnic Institute (UPI-BI-IPN), Mexico

References

[1] Grau, P. (1991). Textile industry wastewater's treatment. Water Sci. Technol. , 24, 97-103.

[2] Beltrán, F. J, Encinar, J. M, & González, J. F. (1997). Industrial wastewater advanced oxidation. Ozone combined with hydrogen peroxide or UV radiation, Water Res. , 31, 2415-2428.

[3] Banerjee, A, & Ghosha, A. K. (2010). Isolation and characterization of hyper phenol tolerant Bacillus sp from oil refinery and exploration sites. J. Hazard. Mater. , 173, 783-788.

[4] Epa, U. S. http://www.epa.gov/waterscience/methods/pollutants.htm, last accessed (2010).

[5] Poznyak, T. I, & Vivero, J. L. (2005). Degradation of aqueous phenol and chlorinated phenols by ozone. Ozone: Sci. Eng. , 27, 447-458.

[6] Saravanan, P, Pakshirajan, K, & Saha, P. (2008). Growth kinetics of an indigenous mixed microbial consortium during phenol degradation in a batch reactor. Bioresour. Technol. , 99, 205-209.

[7] Wei, G, Yu, J, Zhu, Y, Chen, W, & Wang, L. (2008). Characterization of phenol degra-
     dation by Rhizobium sp. CCNWTB 701 isolated from Astragalus chrysopteru in min-
     ing tailing region. J. Hazard. Mater., 151, 111-117.

[8] Abuhamed, T, Bayraktar, E, Mehmetoglu, T, & Mehmetoglu, U. (2004). Kinetics mod-
     el for the growth of Pseudomonas putida F1 during benzene, toluene and phenol bio-
     degradation. Process Biochem., 39, 983-988.

[9] Kumar, A, Kumar, S, & Kumar, S. (2005). Biodegradation kinetics of phenol and cate-
     chol using Pseudomonas putida MTCC 1194. Biochem. Eng. J., 22, 151-159.

[10] Essam, T, Amin, M. A, Tayeb, O. E, Mattiasson, B, & Guieysse, B. (2010). Kinetics and
     metabolic versatility of highly tolerant phenol degrading Alcaligenes strain TW1. J.
     Hazard. Mater., 173, 783-788.

[11] Bajaj, M, Gallert, C, & Winter, J. (2009). Phenol degradation kinetics of an aerobic
     mixed culture. Bioch. Eng. J., 46, 205-209.

[12] Nam, K, & Kukor, J. (2000). Combined ozonation and biodegradation for remedia-
     tion of mixtures of polycyclic aromatic hydrocarbons in soil. Biodegrad., 11, 1-9.

[13] Beltran-heredia, J, Torregrosa, J, Dominguez, J. R, & Garcia, J. (2000). Aerobic biologi-
     cal treatment of black table olive washing wastewaters: effect of an ozonation stage.
     Process Biochem., 35, 1183-1190.

[14] Zhao, G, Zhou, L, Li, Y, Liu, X, Ren, X, & Liu, X. (2009). Enhancement of phenol deg-
     radation using immobilized microorganisms and organic modified montmorillonite
     in a two-phase partitioning bioreactor. J. Hazard. Mater., 169, 402-410.

[15] Chen, K. C, Lin, Y. H, Chen, W. H, & Liu, Y. C. (2002). Degradation of phenol by
     PAA-immobilized Candida tropicalis. Enzyme Microb. Technol., 31, 490-497.

[16] Hong PKAZeng Y. ((2002). Degradation of pentachlorophenol by ozonation and bio-
     degradability of intermediates. Water Res., 36, 4243-4254.

[17] Khokhawala, I. M, & Gogate, P. R. (2010). Degradation of phenol using a combina-
     tion of ultrasonic and UV irradiations at pilot scale operation. Ultrason. Sonochem.,
     17, 833-838.

[18] Chaichanawong, J, Yamamoto, T, & Ohmori, T. (2010). Enhancement effect of carbon
     adsorbent on ozonation of aqueous phenol. J. Hazard. Mater., 175, 673-679.

[19] Derudi, M, Venturini, G, Lombardi, G, Nano, G, & Rota, R. (2007). Biodegradation
     combined with ozone for the remediation of contaminated soils. Eur. J. Soil Biol., 43,
     297-303.

[20] Benitez, F J, Real, F. J, Acero, J. L, Garcia, J, & Sanchez, M. (2003). Kinetics of the ozo-
     nation and aerobic biodegradation of wine vinasses in discontinuous and continuous
     processes. J. Hazard. Mater. B, 101, 203-218.

[21] Aparicio, M. A, Eiroa, M, Kennes, C, & Veiga, M. C. (2007). Combined post-ozona-
      tion and biological treatment of recalcitrant wastewater from a resin-producing fac-
      tory. J. Hazar. Mater. , 143, 285-290.

[22] García-peña, E. I, Zarate-segura, P, Guerra-blanco, P, Poznyak, T, & Chairez, I.
      (2012). Enhanced Phenol and chlorinated phenols removal by combining ozonation
      and biodegradation, , 223, 4047-4064.

[23] Nam, K, Rodríguez, W, & Kukor, J. (2001). Enhanced degradation of polycyclic aro-
      matic hydrocarbons by biodegradation combined with a modified Fenton reaction.
      Chemosphere , 45, 11-20.

[24] Contreras, S, & Rodriguez, M. Al Momani F, Sans C, Esplugas S. ((2003). Contribu-
      tion of the ozonation pre-treatment to the biodegradation of aqueous solutions of 2,4-
      dichlorophenol. Water Res. , 7, 3164-317.

[25] El-Naas, M H, Al-zuhair, S, & Makhlouf, S. (2010). Batch degradation of phenol in a
      spouted bed bioreactor system. J. Ind. Eng. Chem. , 16, 267-272.

[26] Adav, S. S, Chen, M Y, Lee, D. J, & Ren, N. Q. (2007). Degradation phenol by Acineto-
      bacter strain isolated from aerobic granules. Chemosphere , 67, 1566-1572.

[27] Godjevargova, T, Ivanova, D, Aleksieva, Z, & Dimova, N. (2003). Biodegradation of
      toxic organic components from industrial phenol producing wastewater by free and
      immobilized Trichospora cutaneum R 57. Process Biochem. , 38, 915-920.

[28] Edalatmanesh, M, Mehrvar, M, & Dhib, R. (2008). Optimization of phenol degrada-
      tion in a combined photochemical-biological wastewater treatment system. Chem.
      Eng. Res. Des. , 86, 1243-1252.

[29] Dong, Y, Yang, H, He, K, Wu, X, & Zhang, A. (2008). Catalytic activity and stability
      of Y zeolite for phenol degradation in the presence of ozone. Appl. Catal. B: Envi-
      ron. , 82, 163-168.

[30] Ali, M, & Sreekrishna, T. R. (2001). Aquatic toxicity from pulp and paper mill efflu-
      ents: A Review. Advances in Environmental Research , 5, 175-196.

[31] Amat, A. M, Arques, A, Miranda, M. A, & López, F. (2005). Use of ozone and/or UV
      in the treatment of effluents from board paper industry. Chemosphere 60 (8),
      1111-1117.

[32] Bijan, L, & Mohseni, M. (2005). Integrated ozone and biotreatment of pulp mill efflu-
      ent and changes in biodegradability and molecular weight distribution of organic
      compounds. Water Research , 39, 3763-3772.

[33] Industry Profile sponsored by Contaminated Land and Liabilities Division(1996).
      Pulp and Paper Manufacturing Works. Department of the Environment Industry
      Profile, UK

[34] Fengel, D, & Wegener, G. Wood: Chemistry, ultraestructure and reactions". Walter de Gruyter ((1984). Berlin/New York.

[35] Freudenberg, K, & Neish, A. C. (1968). Constitution and Biosynthesis of Lignin. Berlin: Springer-Verlag, 129.

[36] HillCallum A.S. ((2006). Wood modification. Chemical, Thermal and Other Processes. John Wiley & Sons, Ltd, EEUU

[37] AdlerErich. ((1977). Lignin chemistry-past, present and future. Wood Science and Technology, 11(3), 169-218

[38] Demin, V. A, Shereshovets, V, & Monakov, J. B. (1999). Reactivity of Lignin and Problems of its Oxidative Destruction with Peroxy Reagents. Russian Chemical Reviews, 68(11), 937-356.

[39] Peng, G, & Roberts, J. (2000). Solubility And Toxicity Of Resin Acids. Water Research, 34(10), 2779-2785.

[40] RowellRoger. ((2012). Handbook of wood chemistry and wood composites. 2nd Edition. CRC Press Taylor & Francis Group, EEUU.

[41] Ruiz-dueñas, F. J, & Martínez, A. T. (2009). Microbial degradation of lignin: how a bulky recalcitrant polymer is efficiently recycled in nature and how we can take advantage of this. Microbial Biotechnology 2(2), 164-177

[42] Buitron, G, & Gonzalez, A. (1996). Characterization of the microorganisms from an acclimated activated sludge degrading phenolic compounds, Water Sci. Technol. , 34, 289-294.

[43] Kimet., al., ((2002). Biodegradation of phenol and chlorophenols with defined mixed culture in shake-flasks and a packed bed reactor, Process Biochem., , 37, 1367-1373.

[44] Nay, O, Erdeml, E, Kabdali, I, & Lmez, T. (2008). Advanced treatment by chemical oxidation of pulp and paper effluent from a plant manufacturing hardboard from waste paper. Environmental Technology. , 29, 1045-1051.

[45] García-peña, E. I, Zarate-segura, P, Guerra-blanco, P, Poznyak, T, & Chairez, I. (2012). Enhanced phenol and chlorinated phenols removal by combining ozonation and biodegradation, Water, Air, and Soil Pollution, 223 (7), 4047-4064.

[46] Poznyak, T. I, & Vivero, E. J. L. (2005). Degradation of aqueous phenol and chlorinated phenols by ozone, Ozone Science & Engineering, 27 (6), 447- 458.

Chapter 6

Biocomposites: Influence of Matrix Nature and Additives on the Properties and Biodegradation Behaviour

Derval dos Santos Rosa and Denise Maria Lenz

Additional information is available at the end of the chapter

1. Introduction

Composite materials are material systems which consist of one or more discontinuous phases embedded in a continuous phase. Thus, at least two distinct materials that are completely immiscible are combined to form a composite. The continuous phase are termed matrix and the discontinuous phase can be a reinforcement (reinforcing agent) or filler. Also, other additives as plasticizers, pigments, heat and light stabilizers are frequently added in order to provide certain properties. The type and reinforcement geometry impart strength to the matrix and the resultant composite shows optimized properties such as high specific strength, stiffness and hardness with respect to the specific components [1].

As conventional plastics are resistant to biodegradation, the concept of using biobased plastics (biodegradable polymers or biopolymers) as reinforced matrices for biocomposites is gaining more and more approval day by day [2]. A variety of natural and synthetic biodegradable polymers that can be used as biocomposite matrix are commercially avaiable. These biocomposite materials are designed to have a better environmental impact than conventional plastics as well as to promote an improvement in their mechanical properties so that their applications can be expanded. By embedding natural fibers with renewable resource-based biopolymers such as cellulosic plastics; polylactides; starch plastics; polyhydroxyalkanoates (bacterial polyesters); soy-based plastics, the so-called green biocomposites could soon be the future [3].

Biocomposites are composites that present natural reinforcements (like vegetable fibers) in their composition and can be: (i) partial biodegradable with non-biodegradable polymers matrices such as thermoplastic polymers (e.g., polypropylene, polyethylene) and thermoset polymers (e.g., epoxy, polyester) or (ii) fully biodegradable with biodegradable polymers matrices such as renewable biopolymer matrices (e.g., soy plastic, starch plastic, cellulosic plastic) and petrobased biodegradable polymer matrices (e.g., aliphatic co-polyester, polyesteramides). The fully biodegradable ones are 100% biobased materials and show biodegradability and/or compostability properties [2, 4, 5]. For the purpose of this chapter, only fully biodegradable biocomposites are the subject considered.

Natural fiber reinforced plastics by using biodegradable polymers as matrices are the most environmental friendly materials which can be composted at the end of their life cycle. Unfortunately, the overall physical properties of those composites are far away from glass-fiber reinforced thermoplastics. Further, a balance between life performance and biodegradation has to be developed [6].

Hybrid composites are resulted from the incorporation of several types of reinforcing agents with the purpose of tailoring the properties of the obtained composite according to engineering requirements. A synergistic effect between the different kinds of reinforcements enhances the overall performance of the composite. Bionanocomposites are a emerging class of nanostructured biohybrid material which exhibit a singular combination of structural and functional properties together with biocompatibility and biodegradability that was not found in nature. These hybrid materials consist mainly in the assembly of biopolymers and silicates from clay mineral family that have shown extraordinary potential to be used in many applications [7].

In the present chapter, an overview of the current biodegradable polymer matrices and some of the most used reinforcements is described as well as the properties and applications of the obtained biocomposites are dicussed.

2. Biodegradable polymer matrices

There are various ways that biodegradable polymers can be adressed. Depending on their origin, they may be divided as: natural, synthetic or microbial polymers.

2.1. Natural biodegradable polymers

Natural biodegradable polymers are polymers formed naturally during the growth cycle of living organisms. Their synthesis generally involves enzyme-catalyzed reactions and reactions of chain growth from activated monomers which are formed inside the cells by complex metabolic processes. Natural polymers such as proteins (collagen, silk and keratin), carbohydrates (starch, glycogen) are widely used materials for conventional and novel pharmaceutical dosage forms [8]. These materials are chemically inert, nontoxic, less expensive than the synthetic ones, eco-friendly and widely avaiable [8,9]. The families of natural polymers are low-cost materials along with some disavantages such as inferior thermal and mechanical properties. The natural polymers here described are from two groups, i.e., those obtained from vegetable and those from animal sources, as shown in Table 1.

			Cellulose
Plant Source	Carbohydrates	Polysaccharides	Starch
			Pectin
	Proteins		Soy derivatives
			Polypeptides
	Lignins		Polyphenols
Animal Source	Proteins		Silk
			Wool
			Polypeptides
	Polysaccharides		Chitin
			Chitosan
			Glycogen

Table 1. Classification of natural polymers based on their sources.

There are several types of carbohydrates: monosaccharides, disaccharides, oligosaccharides and polysaccharides. The latter ones, of particular interest, are comprised of hundreds or thousands of monosaccharides, commonly glucose, forming linear chains, such as cellulose, or branched chains, as in starch and glycogen. For this chapter, cellulose and its derivatives, starch and chitosan will be presented as natural biodegradable polymers [10].

2.1.1. Cellulose derivatives

Cellulose acetate (CA), universally recognized as the most important organic ester of cellulose because of its extensive applications in fibres, plastics and coatings, is prepared by reacting cellulose with acetic anhydride using acetic acid as a solvent and perchloric acid or sulphuric acid as a catalyst. CA is a carbohydrate composed of β-glucose molecules that are covalently linked through β-1,4-glycosidic bonds, widely found in nature in algae and land plants which has been valued as a functional material. CA comes to meeting the diverse needs of today's society including biodegradability characteristics, its hydrophilic behaviour and biocompatibility [11].

Several applications for cellulose and its derivatives have been shown, for example: in paints, textiles, pharmaceuticals and beauty, fibers, ionic liquids, construction technology and so on [12, 13]. Cellulose esters for coating applications are nearly always used as miscible blends with acrylics, polyesters and other polymers. This is possible because of their ability to form hydrogen bonds through the presence of hydroxyl groups and the carboxyl groups of the ester. An increase in ester molecular weight increases the toughness and melting point but decreases the compatibility and solubility, whereas hardness and density are unaffected. Compatibility, solubility and the maximum non-volatile content all decrease as the ester molecular weight increases. The hydroxyl group content inversely affects the moisture resistance and toughness [11].

Ignácio et al. [14] evaluated the production of cellulosic polymer membranes based on cellulose acetate and thus advanced technology was brougth to be used in membranes for separation processes (ultrafiltration, microfiltration, reverse osmosis, nanofiltration, gas separation, etc.). The use of these membranes has been shown to be effective for water treatment in chemical industries and pharmaceutical processes. Mulinari et al. [15] studied the preparation and characterization of a hybrid composite composed by bleached cellulose and hydrous zirconium oxide. Authors showed that these cellulose composites obtained by the crushed sugarcane combined with an inorganic material has intrinsic advantages such as low cost, biodegradability and simplicity in preparation and handling.

2.1.2. Starch

Starch, a low-cost biodegradable polymer, is abundant in plants, where it is stored in granule form and acts as an energy reserve [16]. Starch is composed of two polymers: amylose and amylopectin, both of which contain α-D-glucose units. Amylose is mostly a linear molecule of $\alpha(1\rightarrow4)$-linked-D-glucopyranosyl units with the ring oxygen atoms all on the same side. Amylopectin is the major branched component of starch and presents a $(1\rightarrow6)$ linkage that forms branch points. The hydrophilicity of these polymers is responsible for their incompatibility with most hydrophobic polymers [17]. When exposed to a soil environment, the starch component is easily consumed by microorganisms, leading to increase its porosity by void formation and the loss of integrity of the plastic matrix. The plastic matrix will be broken down into smaller particles.

Addition of a plasticizer like glycerin can further improve the ductility of starch, forming a polymer that is known as thermoplastic starch (TPS) which is capable of flowing easily. This plastifying agent lowers the glass transition temperature of starch as well as the melting temperature of the mixture by the introduction of mechanical and heat energy. The starch plastification is commonly carried out by extrusion at temperatures close to 120 °C. The mixtures of TPS with other polymers have the potential to behave in a similar manner to more conventional polymer-polymer blends. This would allow greater control of the dispersed phase morphology since the TPS should undergo deformation, disintegration and coalescence [18].

The crystalline nature of starch granules reflects the organization of amylopectin molecules within the granules whereas amylose is the most constituent of the amorphous portion that is randomly distributed among the amylopectin clusters. The conversion of starch into a thermoplastic material by extrusion or by gel casting into films results in the loss of the natural organization of the chains [19]. Figure 1 shows granular starch (a) and pregelatinized starch (b).

Blends of starch with synthetic polymers such as ethylene–vinyl alcohol copolymer, starch/ poly(ethylene-co-vinyl alcohol), copolymers of ethylene with vinyl acetate, vinyl alcohol, acrylic acid, cellulose derivatives and other natural polymers, recycled high density polyethylene (HDPE) and other polyethylenes (PE) as well as compounds with a mixture of glycerin as plasticizer have been studied. Among the environmentally friendly starch-synthetic polymer products currently marketed on a commercial scale are Mater-Bi TM (Novamont, Italy), Bioplast (Biotech, Germany), Biopar (Biop Biopolymer Technologies AG, Germany), and NovonTM (produced by Chisso in Japan and Warner Lambert in the USA [20].

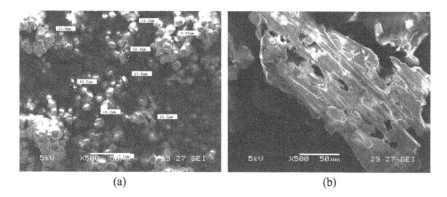

(a)   (b)

Figure 1. Scanning electron microscopy (SEM) photomicrographs of (a) granular starch and (b) pregelatinized starch. Reprinted from Carbohydrate Polymers, 59, Pedroso A. G. and Rosa D. S., Mechanical, thermal and morphological characterization of recycled LDPE/corn starch blends, 1–9, Copyright (2005) [19] with permission from Elsevier.

The blending of biodegradable starch with inert polymers, such as polyethylene (PE), has received considerable attention currently. The reasoning behind this approach is the possibility of disintegration and disappearing of the all plastic films in the waste disposal environment if the biodegradable component is present in sufficient amounts and can be removed by microorganisms.

Pedroso and Rosa [19] studied blends with recycled low density polyethylene (LDPE) and corn starch containing 30, 40 and 50 wt% starch. The blends were prepared by extrusion and characterized by the melt flow index (MFI), tensile test, dynamic mechanical thermal analysis (DMTA) and scanning electron microscopy (SEM). For comparison, virgin LDPE/corn starch blends were prepared and characterized under the same conditions. The addition of starch to LDPE reduced the MFI values, the tensile strength and the elongation at break whereas the modulus increased. The decreases in the MFI and tensile properties were most evident when 40 and 50 wt% were added. SEM images showed that the interfacial interaction was weak for blends containing virgin and recycled LDPE. Blends prepared with recycled LDPE showed the same behavior as those blends prepared with virgin LDPE, indicating that starch was the main factor that influenced the blend.

In other work [21], the same authors blended high density polyethylene (HDPE) and polypropylene (PP), both post-consumer polymers, with thermoplastic starch (TPS). Corn starch plastification was carried out by extrusion with glycerin addition. The processing, thermal and mechanical behaviours of the produced TPS were investigated as well as the morphology characterization of post-consumer HDPE/PP blends (100/0, 75/25, and 0/100 wt.%) in different proportions of TPS (30%, 40% and 50% wt.%). In conclusion, the addition of TPS to recycled PP reduces its melting flow index (MFI) whereas the MFI of HDPE and HDPE/PP blends increases. TPS also decreases the tensile strength and increases the rigidity of the polymers. The incorporation of TPS in polyolefin matrices results in the separation of phases and a disintegration of the starch granules.

2.1.3. Chitosan

Chitosan (CS) is a biopolymer (poly-β-1,4-glucosamine) having immense structural possibilities for chemical and mechanical modifications to generate novel properties, functions and applications, especially in biomedical area. Chitosan is no longer just a waste by-product from the seafood processing industry. This material is now being utilized by industry to solve problems and to improve existing products as well as to create new ones. CS is composed by linear nitrogenous polysaccharides - a basic polysaccharide homopolymer from natural sources, biodegradable, biocompatible and non toxic. Chitosan is produced commercially by deacetylation of chitin, naturally occurring polysaccharides which is the structural element in the exoskeleton of crustaceans (crabs, shrimp, etc.). Due to its variable and incomplete deacetylation process, it acts as a copolymer of varying amounts of N-acetyl glucosamine and N-glucosamine repeated units. The presence of reactive primary amino groups renders special property that makes CS very useful in pharmaceutical applications [22].

CS has three types of reactive functional groups, an amino group as well as both primary and secondary hydroxyl groups. Chemical modifications of these groups have provided numerous useful materials in diferent fields of application. Chitosan oligomers as well as chitosan have been shown to inhibit growth of several fungi and bacteria, especially pathogens. Hirano and Nagao [23] have studied the relationship between the degree of polymerization of chitosan and the inhibition efect.

At room temperature, chitosan forms aldimines and ketimines with aldehydes and ketones, respectively. Reaction with ketoacids followed by reaction with sodium borohydride produces glucans carrying proteic and nonproteic amino groups. N-Carboxymethyl chitosan is obtained from glyoxylic acid and its potential uses are in chromatographic media and metal ion collection [24].

2.2. Biodegradable polymers of microbiological origin

Polymers of microbial origin are produced as intracellular reserve material for a variety of bacteria and have gained prominence due to their possible applications as well as their biodegradable and renewable characteristics.

In the last three decades, the polymers, especially polysaccharides, have acquired great importance in a wide range of industrial processes [25]. Several species of fungi and yeasts produce polymers of commercial interest; however, polymers from bacterial origin are those with greater viability in terms of industrialization and commercialization since they present quality and constant supply. Among these polymers, we highlight the PHB and the PHBV which comprise the group of polyhydroxyalkanoates whose classification is presented in Table 2.

Polysaccharides	Polyhydroxyalkanoates	Poly(3-hydroxy-butyrate) - PHB
		Poly(β-hydroxybutyrate-co-valerate) PHB-V

Table 2. Examples of polyhydroxyalkanoates.

The polyhydroxyalkanoates (PHAs) are thermoplastic polyesters which degrade completely into microbiologically active environments in addition to being biocompatible and may be biosynthesized by a large number of Gram negative and Gram positive bacteria, from different carbon sources or made from renewable and non-renewable genetically modified (GM) plants. Examples of pure cultures used for industrial production of PHAs include *Ralstonia eutropha*, *Alcaligenes lotus*, *Azotobacter vineland* and various Pseudomonas species [26-32].

Genetically modified plants, such as potatoes (*Solanum tuberosum*) and tobacco (*Nicotiana tabacum*) produce cereals such as sunflower and soybean that can provide other ways of producing PHAs. However, the yield (4% of the weight of the plant) is much less than the one obtained by bacteria which reduces the production of PHAs by this method [26-32].

2.2.1. Poly(3-Hydroxy-Butyrate) (PHB)

Poly(3-hydroxy-butyrate), PHB, which is a PHA produced by the *Alcaligens eutrophorus* bacteria, is one of the most interesting biodegradable polymers because it is obtained by bacterial fermentation from renewable resources. PHB can also be synthesized by ring-opening polymerization of β-butyrolactone using distannoxane derivatives as catalysts, such as zinc and alluminium [33]. PHB is linear, homochiral, thermoplastic polyester produced by micro-organisms as intracellular fat deposits in response to limited nutrient availability. PHB belongs to a polyhydroxyalkanoate class of shorter pendant groups that confers a high degree of cristalinity [34].

However, PHB presents some drawbacks like thermal instability at temperatures close to its melting point and a relatively low impact resistance [16]. PHB molar mass decreases proportionately with some processing parameters like time and temperature. In spite of its narrow processing window, PHB with high molar mass can be processed like other thermoplastics if adequate processing parameters are used.

Two main efforts have been used to change PHB properties: biosynthesis and blending. Since blends are a cheaper and faster method to improve polymer properties than synthesis, blends have often been used to improve mechanical properties and processability of PHB. [16, 35].

The biosynthesis of this polymer allows a cyclical process through sustainable renewable sources by replacing cutting edge technologies related to the production and use of synthetic polymeric materials. Among the microorganisms that produce PHB, bacteria like *Alcaligenes eutrophus*, *Azotobacter vinelandii* and *Ralstonia eutropha* can be detached. [36].

According to Lenz et al. [31], the chemical structure of the polyester is an important factor in determining its physical properties and determining the activity of the enzymes involved in their biosynthesis and biodegradation. PHB is a saturated linear polyester, behaving like conventional thermoplastic materials. It has high crystallinity and melting temperature of approximately 176°C. Its glass transition temperature (Tg) is below 5°C and its properties resemble those of polypropylene (PP).

Comparing to polymer commodities, conventional PHB and its copolymers have the advantage of biodegradability and biocompatibility. In contrast, presents the disadvantage of having a poor thermal stability and impact resistance relatively low. Its use spans several segments, such as applications in biomedical areas, agriculture, food packaging and pharmaceutical products, as well as the segments of packaging and agricultural films strongly highlighted. The combination of high temperature and crystallinity provides shine to the films, whereas the rigidity and low impact resistance presented by PHB hinder their use. PHB copolymers have better mechanical properties. The copolymer PHB-V, for example, provide an improvement in ductility and impact resistance, making it more interesting from the point of view of application and end products compared to PHB [30, 32, 37-40].

2.3. Synthetic biodegradable polymers

This class of polymers has been widely used in biomedical uses, such as controlled-release capsules of drugs in living organisms, fasteners surgery (sutures, implants for bone pins) and special packaging. Polymers of this class that have been studied more recently are poly(lactic acid) (PLA), polyglycolic acid (PGA), poly (glycolic acid-lactic acid) (PGLA) and poly(e-caprolactone) (PCL) [35].

For greater understanding, synthetic biodegradable polymers are separated into classes. Table 3 shows the classification of non-natural synthetic biodegradable polymers.

	- PLA
Aliphatic	Poly(glycolic acid) - PGA
	Poly(e-caprolactone) - PCL
	Polytrimethylene terephthalate - PTT
Aliphatic Aromatics (PAA)	Poly(butylene terephthalate) -PBT
	Poly (butylene succinate) - PBS

Table 3. Classification of non-natural synthetic biodegradable polymers.

The polyesters compete an important position among the group of biodegradable plastics and some biodegradable polyesters are already commercially available.The main biodegradable polyesters are those based on hydroxy-carbonic acids. The biodegradable polyesters still have high cost, but they have aroused great interest due to their accessible production by fermentation or synthetic routes [35].

During the last two decades, aliphatic polyesters such as poly(ε-caprolactone) (PCL) and poly (L-lactic acid) (PLLA) have been extensively studied due to their ability to undergo hydrolysis in the human body as well as in natural circumstances [37, 41, 42].

2.3.1. Poly(Lactic Acid) (PLA)

Poly(lactic acid) (PLA) is a hydrolytically degradable aliphatic polyester which presents water vapor permeability that may have a significant influence on its rate of degradation. The poly(lactic acid) (PLA) is an aliphatic polyester obtained by polymerization of lactic acid. This can be found in the form of two optical isomers: L-and D-lactide. PLA has potential for applications in the medical, pharmaceutical and packaging, mainly as implantable devices temporarily (sutures, staples, nano-reservoirs for drugs etc). Other applications involve the sectors of textiles and fibers, agriculture, electronics, appliances and housewares [43, 44, 45].

PLA presents some advantages like biocompatibility, has better thermal processibility compared to other biopolymers such as poly(hydroxyalkanoates) (PHAs), shows eco-friendly characteristics and requires 25–55% less energy to produce than petroleum-based polymers. Nevertheless, PLA is a very brittle material and chemically inert with no reactive side-chain groups making its surface and bulk modifications a challenging task. Besides, PLA shows low degradation rates and is hydrophobic [46].

Henry et al. [47] investigated systems including poly (lactic acid) (PLA). The thermal analysis showed that the glass transition temperature (Tg) of the polymer is about 320 K. The β relaxation was observed between -150 °C and -30 °C, depending on the measurement frequency (1 Hz - 100 kHz) and was determined as secondary relaxation in the glassy state. The authors studied the changes that are associated with water penetration into the polymer which directly affect the relaxation process. Water molecules confined (outlined / permeated) and the polymer chains in polymer networks represent an important function in matrix degradation and, thus, the authors were able to observe the evolution of degradation for a few weeks in an environment with controlled humidity. It is accepted that water penetrates preferentially in amorphous areas, but also affects the crystalline regions. It is a clear evolution of the observed activation energy of relaxation during polymer degradation. The resulting dielectric relaxations are complemented with measures of molecular weight during degradation with time.

2.3.2. PCL

Poly (e-caprolactone) (PCL) is a synthetic aliphatic polyester made from ring opening polymerization. This biodegradable polyester presents good mechanical properties that is compatible with many types of polymers and is one of the most hydrophobic biodegradable polymers currently available. PCL has been widely studied for use in drug release systems [48]. Extracellular enzymes present in soil can cleave the extensive chains of PCL before assimilation of the polymer by microorganisms. However, the high cost of PCL has prevented its widespread industrial use. PCL has been thoroughly examined as a biodegradable medium and as a matrix in controlled drug-release systems [14, 49].

The main limitation of PCL is its low melting temperature (Tm 65°C) and also has some drawbacks, including a poor, long-term stability caused by water absorption, poor mechanical and processing properties. Some of these problems can be overcome by physical or chemical modifications, including the blending of these polymers. [49]

PCL/CA blends are generally incompatible, immiscible and show poor interpolymeric adhesion [14, 49]. Rosa et al. [11] reported miscibility between several CAs and aliphatic polyesters. The miscibility of the cellulose polymer with a polymeric plasticizer is important in order to maintain the already complex mixture as homogenous as possible. The use of coupling agents usually improves the elongation of composites, but frequently results in a decrease in strength. One approach to improve the compatibility between the constituent polymers in PCL/CA mixtures is to incorporate a compatibilizer into the mixture. The chemical modification of aliphatic polyesters by grafting is another way of improving the compatibility between starch and aliphatic polyesters in polymeric blends. The effects of polyethylene grafted with maleic anhydride (PE-g-MA) on the thermal and mechanical properties, as well as on the morphology of blends of low-density polyethylene (LDPE) and corn starch have been studied using differential scanning calorimetry (DSC), tensile strength measurements and scanning electron microscopy [14, 49-53].

3. Natural reinforcement agents as additives for biocomposites

Polymer reinforcements are generally used in order to provide stiffness and strength to the polymer matrix resulting in improved mechanical properties for the obtained composites Besides, properties like water and gas barrier as well as fire resistance and flame retardant properties and so on can be enhanced by the employ of reinforcements in polymer matrices [54-56].

The present review focuses on vegetable fibers (also reported as natural or plant fibers), nanofibers extracted from them and nanoclays in particular mineral silicates as reinforcement agents for biobased polymer matrices. Instead of being a natural non-renewable source, nanoclays are abundantly available and improve mechanical properties at lower loadings [57].

3.1. Natural or vegetable fibers

The interest in the use of vegetable fibers as reinforcement agents in polymeric composites is growing currently owing to environmental regulations and ecological concerns of the actual society.

Vegetable fibers are abundantly available, fully and easily recyclable, non-toxic, biodegradable, non-abrasive to the molding machinery, easily colored as well as have lower cost, lower density and lower energy consumption in producing step with respect to synthetic fibers as glass and carbon fibers [58,59]. In addition to having lower processing energy requirements and more shatter resistant when compared with synthetic fibers, vegetable fibers have good sound abatement capability, non-brittle fracture on impact, high specific tensile modulus and tensile strength, low thermal expansion coefficient and low mold shrinkage [59].

There are thousands of different fibers in the world and a few of them have been studied. All vegetable fibers (wood or non-wood fibers) are constituted by cellulose; hemicellulose and lignin combined to some extent as major constituents [6]. In fact, the so-called lignocellulosic fibers have cellulose as the main fraction of the fibers. Cellulose is a semicrystalline polysaccharide made up of D-glucosidic bonds. A large amount of hydroxyl groups in cellulose (three in each repeating unit) imparts hydrophilic properties to the natural fibers [60]. Thus, they are hydrophilic in nature. Cellulose forms slender rodlike crystalline microfibrils that are embedded in a network of hemicellulose and lignin, i. e., the microfibrils are bonded together through an amorphous and complex lignin/hemicellulose matrix that acts as a cementing material.

Hemicellulose is a polysaccharide with lower molecular weight than cellulose.The main difference between cellulose and hemicellulose is that hemicellulose has much shorter chains and also has branches with short lateral chains consisting of different sugars while cellulose is a linear macromolecule [52]. Both are easily hydrolyzed by acids, but only hemicellulose is soluble in alkali solutions as well as lignin. Lignin is a hydrocarbon polymer with a complex composition that presents hydroxyl, methoxyl and carbonyl functional groups [4].

Lignocellulosic fibers may be found in different parts of the plant like leaf, bast, seed and fruit. Some fibers derived from leaf part - leaf fibers: abaca (Manila hemp), sisal, curauá, banana leaf fiber, pineapple leaf fiber (PALF) and henequen; fibers derived from the inner bark part - bast fibers: flax, ramie, kenaf/mesta, hemp, piaçava and jute; fibers derived from plant seed - seed fibers: cotton and kapok and fruit fibers: coconut husk, i.e., coir and luffa. Climatic conditions, age of plant and the digestion process influence not only the structure of fibers but also their chemical composition [56, 61]. Plant fibers from wheat straw, rice straw, oat straw, esparto, elephant grass, bamboo, bagasse (sugar cane) are classified as grass and reed fibers [56] Some of these non-wood fibers were been studied as raw material source (pulp) for papermaking in many developing countries and for biocomposites manufacture whose composites can be applied mainly for food or non-food packaging, automobile parts and biomedical engineering in repairing or restoring tissues and implants as well as drug/gene delivery [62, 63].

Wood fibers have numerous types distributed in softwoods and hardwoods. Hardwoods are, in general, more complex and heterogeneous in structure than softwoods having a character-istic type of cell called vessel element (or pore) for water transport [64].

Table 4 shows the chemical composition of some non-wood vegetable fibers. The concentration of cellulose and other components of lignocellulosic fibers exhibit a considerable variation even for the same fiber. The references therein indicate concentration values all along the presented concentration range. The spiral angle of the cellulose microfibrils and the content of cellulose, determines generally the mechanical properties of the cellulose-based natural fibers [6]. For instance, these two structure parameters were used to calculate the Young's modulus of the fibers through models developed by Hearle et al [65] cited by Bledzski and Gassan [6].

As natural materials, vegetable fibers have nonuniformity such in dimensions as in mechanical properties when compared to synthetic fibers. Other drawbacks for the use of vegetable fibers in biocomposites are: (i) the lower processing temperature (limited to approximately 200°C) due to fiber degradation and/or volatile emissions; (ii) the high moisture absorption due to fiber hydrophilic nature and (iii) incompatibility with most hydrophobic polymers. These problems are well known and countless research has been developed to reduce them with reasonable success [66, 67]. Nevertheless, vegetable fibers (as fillers or reinforcements) are the latest growing type of polymer additives [68].

Because of the low interfacial properties between vegetable fiber and polymer matrix which often reduce their potential as reinforcing agents due to fiber hydrophilic nature, chemical modifications are considered to optimize the interface of fibers. Chemicals may activate hydroxyl groups or introduce new moieties that can effectively interlock with the matrix [69].

Fiber	Chemical Composition					
	Cellulose (wt%)	Lignin (wt%)	Hemicellulose (wt%)	Ash (wt%)	Microfibrilar/spiral angle (Deg.)	References
Abaca	56-63	7-13	15-25	5	--------	[2, 56, 68, 69]
Curauá	70.7–73.6	7.5-11.1	9.9	0.9	--------	[2, 66, 67]
Flax	64-71	2–5	18.6–20.6	5	5-10	[2, 56, 68, 69]
Hemp	57-77	3.7-13	14-22.4	--------	2-6.2	[2, 56, 68, 69]
Henequen	77.6	13.1	4-8	--------	--------	[68, 69]
Jute	45–72	12–26	12–21	0.5–2	8.0	[2, 6, 56, 68, 69]
Kenaf	31–72	8–21	22–24	2–5	--------	[2, 56, 68, 69]
PALF	70-82	5-12.7	--------	--------	14	[2, 68]
Ramie	68.6–91	0.6–0.7	5–16.7	--------	7.5	[2, 6, 56, 68, 69]
Sisal	47–78	8–13	10–24	0.6–1	10-22	[2, 6, 56, 68, 69]

Table 4. Chemical composition of some common vegetable fibers.

Over the last decade, many approaches towards enhancing interfacial adhesion have been pursued. Generally improvements can be accomplished, but there must be a critical cost-benefit evaluation of using the added interfacial agents or processing steps [63].

Alkaline treatment or mercerization is one of the most used chemical treatments of natural fiber. The important modification done is the disruption of hydrogen bonding in the fiber network structure, increasing surface roughness. This treatment removes a certain amount of lignin, wax and oils covering the external surface of the fiber cell wall, depolymerizes cellulose and exposes the short length crystallites [69, 70]. As a result; the adhesive characteristics of the fiber surface are enhanced [71]. Figure 2 shows the aspect of curauá vegetable fiber before and after treatment of NaOH solution.

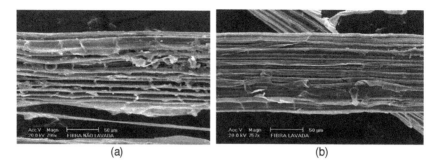

(a)   (b)

Figure 2. SEM micrographs of curauá fiber: (a) as received (b) washed with 0.1 M NaOH solution 24 h at room temperature. Source: Authors

The efficiency of the alkali treatment depends on the type and concentration of the alkaline solution as well as time and temperature of the treatment. Different conditions for alkali treatment of vegetable fibers can be found in literature as well as combinations with other treatments [6, 72].

Authors reported that alkali concentration and reaction time of mercerization has a significant effect on the surface modification [73]. C. indica vegetable fibers were immersed firstly in 2% NaOH for the different time intervals at room temperature to optimize the mercerization time. Afterwards, the mercerization of C. indica fiber was also carried out in 4, 6, 8, 10, 12, and 14% NaOH solutions to study the effect of different concentrations of NaOH on the mercerization of the fibers. Maximum mercerization observed in terms of weight loss of fiber polymer backbone was observed at 210 min. With respect to the concentration of NaOH solution, the weight loss increases with the increase in alkali concentration and shows maximum weight loss at 10% alkali concentration. This happens due to the removal of lignin, hemicelluloses, pectin and other surface impurities with NaOH.

Campos et al. [74] reported the development of biocomposites of thermoplastic starch and polycaprolactone (PCL) with sisal fibers as reinforcement agent. Sisal fibers were treated with sodium hydroxide solution (NaOH 5% (w/v) at 90ºC under agitation for 60 min. After that, sisal fibers were bleached with a blend solution of peroxide hydrogen (H_2O_2 16%) and sodium hydroxide (NaOH 5%) at 55 ºC for 90 min. The authors observed strong adhesion fiber-matrix and interaction between carboxyl groups in PCL-starch and hydroxyl groups in sisal fibers.

Nevertheless, alkaline treatment or other chemical/physical treatment may damage vegetable fiber surface structure, reducing its strength [75, 76]. When a chemical treatment is applied on synthetic fibers like glass fibers only fiber surface is modified. On the contrary, chemical treatments applied on vegetable fibers can produce important chemical and structural changes not only at fiber surface but also on the interphase between elementary fibers [66]. Furthermore, the orientation of microfibrils of cellulose within each elementary fiber plays an important role because it changes the crystallinity of the natural fiber [77]. A different variety of chemical treatments applied on sisal fibers resulted in greater extensibility and lower modulus. These phenomena must be related to the structural variation in the ultimate cells, that is, swelling and partial removal of lignin and hemicellulose [78].

Moraes et al. [76] showed the use of sodium borohydride ($NaBH_4$) (1% wt/vol) as protective agent for vegetable sisal fibers under alkaline treatment with sodium hydroxide (NaOH). The authors reported that the effectiveness of hydride ions (H^-) to protect the sisal fiber was more pronounced in moderate NaOH concentrations (5 wt/vol %) at room temperature or higher (10 wt/vol %) for shorter alkaline treatment times.

Acetylation of natural fibers is a well-known esterification method causing plasticization of cellulosic fibers. Acetylation reduces the hygroscopic nature of natural fibers and increases the dimensional stability of composites [54]. Acetylation is based on the reaction of cell wall hydroxyl groups of lignocellulosic materials with acetic or propionic anhydride at elevated temperature [70]. Other chemical treatments that have already used for fiber treatment are mainly benzoylation treatment, permanganate treatment, isocyanate treatment and peroxide treatment [69].

The use of coupling agents is also extensively used for chemical modification of synthetic and vegetable fibers. Organosilanes and maleic anhydride are both coupling agents that not only produce surface modification but also can produce grafting polymers [63, 79]. Acrylonitrile grafting has also been reported as fiber treatment for glass fibers as well as for vegetable fibers [69]. Coupling agents can be found inserted in polymer matrices (grafted polymer matrices) or in vegetable fibers or even introduced during reactive melt processing of the biocomposite.

In work of Chang et al. [80], kenaf fiber dust was added to a previous maleated polycaprolactone/thermoplastic sago starch blend used as biocomposite matrix. The addition of Kenaf fiber up to 30 phr decreased the water absorption capacity of the maleated treated biocomposites with respect to non-treated biocomposites. The decrease in water absorption was due to the enhanced adhesion between the Kenaf fiber dust and the matrix through grafting which led to decrease of voids between fiber/matrix interfaces. Besides, Kenaf fiber addition improved the mechanical properties of the maleated and non-maleated biocomposites. Nevertheless, tensile strength and modulus reached higher values for maleated biocomposites with higher Kenaf fiber loadings. The effective coupling mechanism of maleic anhydride between polymer matrix and Kenaf has been attributed to esterification reaction between the hydroxyl groups of the Kenaf and anhydride group to form ester linkages [69, 80].

Different authors have applied different methods for silane treatment and have studied the effect of silane treatment on surface morphological and hygroscopic character of the natural fibers. Most of the silane groups have the following formula: $R_{(4-n)} - Si - (R'X)_n$ ($n = 1,2$) where R is alkoxy, X represents an organofunctionality, and R' is an alkyl bridge connecting the silicon atom and the organofunctionality [81].

Some authors prepared bamboo fiber-reinforced polylactic acid (PLA) biocomposites using a film-stacking process [71]. Bamboo fibers were subjected to three different silane treatments: direct silane coupling, silane coupling after plasma treatment and silane coupling during UV irradiation. Biocomposites with silane coupling after plasma-treated fibers presented the highest increase in tensile strength with respect to biocomposites with untreated fibers and among all tested fiber treatments, showing a close adhesion between the PLA matrix and fibers. Fiber surface modifications was related to the silane that should have two functional groups to effectively couple fiber and matrix: a hydrolyzable alkoxy group to condense with hydroxyls on the surface of bamboo fibers and an organofunctional group capable of interacting with the PLA matrix that can result in a copolymerization (grafting) and/or formation of a interpenetrating network.

Other works [81, 82] also reported that in general the interaction of the silane coupling agent with vegetable fibers involves four steps: (i) hydrolysis of silane monomers in presence of water to yield reactive silanol (–Si-OH), (ii) self-condensation of silanol, (iii) The silanol monomers or oligomers are physically adsorbed to hydroxyl groups of fibers by hydrogen bonds on the fiber surfaces and/or in the cell walls. The free silanols also adsorb and may react with each other forming rigid polysiloxane structures linked with a stable –Si-O-Si– bond and (iv) grafting under heating conditions since the hydrogen bonds between the silanols and the hydroxyl groups of fibers can be converted into the covalent –Si-O-C– bonds and liberating water.

In order to enhance the behavior of Kenaf/PLA biocomposites, authors [43] treated kenaf fibers with sodium hydroxide and 3-aminopropyltriethoxysilane (APS) coupling agent. The authors described the hypothetical reaction of silanol and the fiber: the ethoxy groups of APS hydrolyze in water or a solvent producing a silanol and next the silanol reacts with the OH group of the kenaf fiber which forms stable covalent bonds to the cell wall that are chemisorbed onto the fiber surface. In other work [83], ramie fibers were treated with permanganate acetone solution and with permanganate acetone solution followed by silane acetone solution to produce biocomposites with poly(L-lactic acid) PLLA matrix by hot press molding. The fiber surface-treatment with permanganate acetone solution followed by silane acetone solution improved the interfacial adhesion with PLLA matrix. Both treatments accelerate the water permeation rate in PLLA biocomposites, which plays a critical role in the decline of interfacial adhesion strength.

Also, physical treatments have been used. These treatments change structural and surface properties of the fiber and thus influence the mechanical bonding with the polymer matrix. Some pf these treatments envolve fibrillation and electric discharge (Corona, cold plasma, sputtering) and so on [72]. Cold plasma treatment causes chemical implantion, etching, polymerization, free radical formation and crystallization whereas sputtering promotes physical changes such as fiber surface roughness that leads to fiber/matrix interface adhesion [71, 84].

Nevertheless, the hydrophilic character of natural (biobased) polymers has contributed to the successful development of environmentally friendly composites, as most natural fibers and nanoclays are also hydrophilic in nature [85]. Most of the published studies on biocomposites with biodegradable polymers are with polyester matrix, such PHA, due to its polar character that provides better adhesion to lignocellusic fibers [86].

Authors [87] showed that curauá vegetable fibers have good interfacial adhesion to a polyester-based matrix even without coupling agent addition. In this work coupling agent was added during reactive extrusion at the same time with the neat matrix and a masterbatch containing curauá fiber and the blend matrix. The authors reported the importance of the coupling agent addition, beside the NaOH treatment of the fiber, for improved interfacial fiber/matrix adhesion. Figure 3 shows SEM analysis of tensile fracture cross-section samples of polyester blend/curauá fiber biocomposite. Figure 3a revealed a weak fiber/matrix interface with numerous irregularly shaped microvoids and some de-bondings for composites in the absence of coupling agent, which could be responsible for deterioration of the stress transfer from the matrix to the fibers having an adverse effect on the mechanical properties. On the other hand, composites with coupling agent showed an improvement in polymer/fiber adhesion, avoiding fiber pull-out that leads to voids emerging. In this case, curauá fibers were broken under tension (Figure 3b).

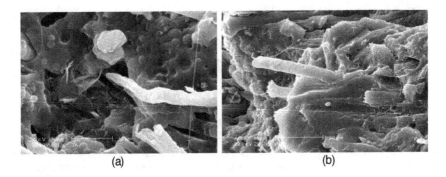

(a)   (b)

Figure 3. SEM micrographs of fracture cross section of polyester blend/curauá fibers: (a) without coupling agent and (b) with coupling agent. Reprinted with kind permission from Springer Science and Business Media: Journal of Polymers and the Environment Biodegradable Polyester-Based Blend Reinforced with Curauá Fiber: Thermal, Mechanical and Biodegradation Behaviour 20, 2012, 237-244, Harnnecker F., Rosa, D. S., Lenz, D. M., Figure 3a and 3b [87].

3.2. Cellulose nanofibers from vegetable fibers

Cellulose is the most abundant renewable carbon resource on Earth. Thus, it can be obtained from many natural sources. Aside from occurring in wood, cotton and other plant-based materials derived from agricultural crops and by-products, cellulose is also produced by algae, some bacteria and tunics of marine animals – tunicates. [88, 89]. The main difference between cellulose obtained by plants and bacteria is that plant-synthesised cellulose usually also contains hemicellulose, lignin and pectin while cellulose produced by bacteria on the other hand, is pure cellulose without foreign substances [90]. Also, highly crystalline cellulose in the native state can be extracted from tunicates which shows high aspect ratio (length/diameter ratio) as well as allows better matrix-to-filler stress transfer [91].

Nanofibers are fibers that have at least one of its linear dimensions smaller than 100 nm. One of the more significant characteristics of nanofibers is the enormous availability of surface area per unit mass - 1 m^2 of them weighs only 0.1 - 1 gram [3, 92]. Cellulose nanofibers are one class of natural fibers that have resulted in structures with remarkable mechanical properties. These nanofibers have received an increasing interest in the bio-based materials community since nanocellulose reinforced biopolymers will be less expensive than many common plastics derived from petroleum resources if processing costs can be kept to between \$0.20–\$0.25/lb [93]. However, the full reinforcing potential of nanofibers has yet to be realized partly because of issues related to scaling manufacturing processes [94].

Cellulose nanofibers are nano-reinforcements from biomass that have been improved the biobased polymers properties such as thermal stability, mechanical toughness and barrier properties at much lower fiber fractions than those required in conventional vegetable fiber composites. Biocomposite materials have been showed potential to be used in packaging with PLA matrix [95] and medical applications using polyurethane - PU - matrix [96].

There are many different methods to obtain nanofibres from vegetable fibres. Cellulose nanocrystals, also reported in the literature as nanowhiskers (or just simply "whiskers"), nanofibers, cellulose crystallites or crystals, are the crystalline domains of cellulosic fibers, isolated mainly by acid hydrolysis [97].

Cellulosic materials intended for use as nano-reinforcements in biocomposites are usually subjected to hydrolysis by strong acids such as sulfuric or hydrochloric acid, yielding in a selective degradation of amorphous regions of cellulose and, consequently, the splitting of micro-fibril beams. As a result of cellulose hydrolysis, the disintegration of its hierarchical structure takes place to form crystalline nanofibers [89]. Usually the acid hydrolysis is combined with sonication [88]. The source of cellulose and hydrolysis conditions (acid concentration, acid to cellulose ratio, temperature and reaction time directly affect the morphology of the nanocrystals [89, 98]. The length of the so-produced nanocrystals generally ranges between 100 and 300 nm and width of 5-20 nm [88, 99]. Invariably these nanocrystals from plant fibers present a rod-like structure [91].

Cellulose nanoparticles are obtained as stable aqueous suspensions and thus the processing of cellulose nanocomposites was first limited to using hydrosoluble (or at least hydrodispersible) or latex-form polymers as nanocomposite matrices. After dissolution of the hydrosoluble (or hydrodispersible) polymer, the aqueous solution was mixed with the aqueous suspension of cellulosic nanoparticles to form a mixture that was cast and evaporated to obtain a solid nanocomposite film. The use of the extrusion processing technique was hampered due to the hydrophilic nature of cellulose which causes irreversible agglomeration of the nanofibers in polymer matrices [3]. The development of newer industrially viable processing techniques as melt compounding is the focus currently. PLA nanocomposites reinforced by cellulose nanofibers separated from kenaf pulp were obtained using a two-step process: masterbatch preparation using a solvent mixture of acetone and chloroform followed by extrusion process and injection molding. The tensile modulus and the tensile strength of the PLA nanocomposite using 5 wt% of nanofiber showed an increase of 24% and 21%, respectively [100].

Cellulose nanocrystals can also be produced by submitting vegetable fibres to high mechanical shearing forces, disintegration of the fibres occurs, leading to a material called microfibrillated cellulose (MFC) [88, 101]. However, depending upon the raw material and the degree of processing, chemical treatments (alkaline, enzimatic or oxidation treatments) may be applied prior to mechanical fibrillation which aim to produce purified cellulose, such as bleached cellulose pulp, which can then be further processed [101]. These nanofibrils ideally consist of individual nanoparticles with a lateral dimension around 5 nm, but MFC generally consists of nanofibril aggregates, whose lateral dimensions range between 10 and 30 nm or more [88].

The major obstacle when producing cellulose based nanocomposites is to disperse the hydrophilic reinforcement in the hydrophobic polymer matrix without degradation of the biopolymer or the reinforcing phase. This can be addressed by improving the interaction between cellulose nanofibers and the matrix and/or by using suitable processing methods [102]. Jute nanofibers submitted to alkali, dimethyl sulfoxide (DMSO) and acid hydrolysis treatments were incorporated into the biocopolyester matrix by melt mixing in varying weight percentages ranging from 0% to 15%. The enhancement in properties was highest for 10 wt % jute nanofiber loaded composites, indicating the most uniform dispersion in this material [103]. In work of Wang and Drzal [104], the solvent evaporation technique (commonly used for drug microencapsulation) was employed to suspend PLA in water as microparticles. The suspension of the PLA microparticles was mixed with high pressure homogenized cellulose nanofibers, producing nanocomposites with good fiber dispersion after water removal by membrane filtration followed by compression molding. Tensile modulus and strength increased up to 58% and 210%, respectively, with respect to neat PLA.

In other work, a hybrid multi-scale biocomposite composed by microfibrillated cellulose (MFC) and bamboo fiber bundles in a polylactic acid (PLA) matrix were successfully processed by extrusion using a surfactant which favoured the dispersion of nanowhiskers in PLA matrix [105]. A hierarchy structure of reinforcement was created with bamboo fiber as the primary reinforcement and cellulose creates an interphase in the PLA matrix around the bamboo fiber that prevents sudden crack growth.

In work of Cherian et al. [106], the nanodimensional cellulose embedded in pineapple fibers was extracted applying acid coupled steam treatment. This treatment was found to be effective in the depolymerization and defibrillation of the fiber to produce nanofibrils of these fibers. Figure 4 shows the cellulose nanofibers extracted through this treatment. These nanofibrils were used to reinforce the polyurethane (PU) by compression moulding [96]. The addition of 5 wt% of cellulose nanofibrils to PU increased the strength nearly 300% and the stiffness by 2600%. The developed composites were utilized to fabricate various versatile medical implants.

A new type of modification of vegetable fibers which consists in the deposition of a nanosized cellulose coating onto natural fibers or the dispersion of nanosized cellulose in natural fiber reinforced composites has been studied in order to develop hierarchical structures. This fiber modification has great potential to improve the fiber-matrix interface and the overall mechanical performances of such composites. Nevertheless, the aspect ratio and alignment of the cellulose nanofiller need optimization as well as novel processing techniques need to be developed to take advantage of the potential use of cellulose nanocrystals [107].

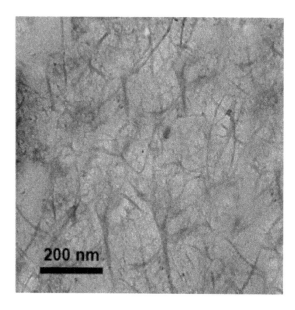

Figure 4. Transmission electron micrograph of cellulose nanofibers from pineapple fibers. Reprinted from Carbohydrate Polymers, 81, Bibin Mathew Cherian, Alcides Lopes Leão, Sivoney Ferreira de Souza, Sabu Thomas, Laly A. Pothan, M. Kottaisamy Isolation of nanocellulose from pineapple leaf fibers by steam explosion, 720–725, Copyright (2010) [106] with permission from Elsevier.

3.3. Nanoclays

Various inorganic nano-particles have been recognized as possible additives to enhance the polymer performance such as polymer nanofibers, the cellulose whiskers and the carbon nanotube. Among these, up to now only the layered inorganic solids like nanoclay have attracted some attention by the packaging industry. This is not only due to their availability and low cost but also due to their relative simple processability and significant improvements in some properties of the resulting polymer composites that include [108, 109]:

• Mechanical properties;

• Decreased permeability to gases, water and hydrocarbons;

• Thermal stability and heat distortion temperature;

• Flame retardancy and reduced smoke emissions;

• Chemical resistance;

• Surface appearance;

• Electrical and thermal conductivity;

• Optical clarity in comparison to conventionally filled polymers.

Most of synthetic bionanocomposites result from the assembly of biopolymers and silicates belonging to the clay mineral family. The effect of nanoclay minerals on polymer properties is mainly attributed to their high surface area and high aspect ratio as well as the combination of singular properties such as chemical inertness, low or null toxicity, good biocompatibility with high adsorption ability and cation exchange capacity [110]. Nanoreinforcement of biobased polymers with nanoclays can thus create new value-added applications of "green" polymers in the materials world [111].

Montmorillonite (MMT) clays, part of the smectite family clays, are the clay minerals most used as fillers in polymer nanocomposites due to environmental and economic criteria [112]. The chemical structure of MMT clays consist of two fused silica tetrahedral sheets sandwiching an edge-shared octahedral sheet of either magnesium or aluminum hydroxide establishing a nanometer scale platelets of magnesium aluminum silicate [113]. Each platelet of MMT is about 1 nm in thickness and varies in lateral dimension from 50 nm to several micrometers, showing high aspect ratio. Also, the platelet has a negative charge arising from isomorphous substitution in the lattice structure, which is compensated by naturally occurring cations that are located within the gallery (or interlayer) regions between the platelets [8]. Clay structure is formed by hundreds of layered platelets stacked into particles or tactoids of approximately 8 to 10 μm in diameter [114, 115].

MMT clays have hydrophilic nature due to the presence of inorganic cations on the basal planar surface of montmorillite layer [116]. The hydrophilicity of the surface of MMT clays makes their dispersion in organic matrices difficult [117]. Thus, MMT clays must be submitted to treatments which play an important role in the preparation of nanocomposites since it can affect their final properties. The most widely used treatments are the diverse functionalizations of clay by various organic cations through ion exchange where the inorganic cations are replaced by organic cations intercalated into the silicate layers. Its hydrophilic nature and ionic exchange capacity allow the silicate mineral to be intercalated by organic cations, which in most cases are alkylammonium ions, to make the clay organophilic and compatible with polymer matrices, preferably with polymers with polar groups which exhibit a higher affinity towards the alkylammonium ion-modified clays [118]. Functionalization of MMT clay by means of the silylation reaction with 3-aminopropyltriethoxysilane and N-[3-(trimethoxysilyl)propyl]ethylene-diamine was also reported [119].

There are three possible morphologies for polymer-clay nanocomposites that include: (i) immiscible, (ii) intercalated and (iii) exfoliated structures [115, 120]. In the immiscible structure the polymer does not penetrate between the clay platelets and the interlayer space of the clay gallery does not expand due to its poor affinity with the polymer, so this structure is also known as phase separated morphology or tactoid morphology. Intercalation is attained when polymer chains slightly penetrate within the gallery space and induce moderate expansion of the clay platelets. Exfoliation is characterized by a random distribution of the clay platelets due to extensive penetration of the polymer chains, resulting in the delamination of the clay platelets and the loss of the crystalline structure of the clay. This is due to a high affinity between polymer and clay.

There are three main processing routes for the development of well dispersed clay/biobased nanocomposites [108, 121]: (i) the solvent route which consists in swelling the layered silicates in a polymer solvent, (ii) the *in-situ* polymerization route for which the layered silicates are swollen in the monomer or monomer solution so as the polymer formation can occur between the intercalated sheets and (iii) the melt processing route which is based on polymer processing in the molten state (extrusion, injection molding, etc) which is highly preferred in the context of sustainable development since it avoids the use of organic solvents.

4. Biocomposites of biobased polymers and natural reinforcement agents: Properties and applications

The development of biocomposites started in the late 1980s and most of the biodegradable polymers which are now available in the market do not yet satisfy each of the requirements for bio-composites. Although promising results were obtained, development of biocomposites is still in its preliminary stage. More data on properties of biocomposites are required to establish confidence in their use [122]. Nanotechnologies promise many stimulating changes in composite materials in order to enhance health, wealth and quality of life, while reducing the environmental impact [108]. Thus, many researches in the biocomposite area can be found in literature. Some of them are reported in the following items.

4.1. PLA based biocomposites

One of the most studied biocomposites is PLA (polylactide) based biocomposite since PLA was the first commodity plastic produced from annually renewable resources [123]. Lactid acid based polymers (polylactides) are polyesters made from lactic acid. PLLA (poly-L-lactide) is a polymer built with only repeating units of L-stereoisomer configuration. The general term PLA (polylactide) is used for polymers without isomer specification.

PLA is brittle, so it needs modification for pratical applications. Bledzki and Jaszkiewicz [124] reported that one of the main drawbacks concerning technical applications of biodegradable polymers, especially for PLA polymers, is their low impact strength. Most research on PLA biocomposite ultimately seeks to improve the mechanical properties to a level that satisfies a particular application [125]. The mechanical properties of biocomposites depend on a number of parameters such as percentage of fiber content, interfacial characteristics between fiber and matrix, fiber aspect ratio, surface modification of fibers and addition of various additives (coupling agents) to enhance the compatibility between fiber and matrix [126].

Huda et al. [82] studied the addition of alkali and/or silane treated Kenaf fibers in PLA matrix through compression molding using the film-stacking method with a fiber content of 40 wt%. Although the introduction of treated kenaf fibers significantly improves flexural modulus compared to the neat PLA matrix, the flexural strength of the PLA composites decreases with the addition of Kenaf fibers. The composite with silane-treated fibers showed an increase of 69% in modulus than that of alkali treated fibers. The notched Izod impact strength of surface-treated composites was higher than those of the neat PLA. The impact strength of neat PLA improved almost 45% with the addition of 40 wt% untreated fiber and 90% with alkali treated Kenaf fibers with the same content. The high toughness of this natural fiber laminated biocomposite places it in the category of tough engineering materials. Other authors [63] used a carding process that provided a uniform blend of PLA fiber and Kenaf fiber that was followed by needle punching, pre-pressing and further hot-pressing in presence of silane coupling agent to form the biocomposite material. The flexural modulus and flexural strength of the treated fiber biocomposites increased with respect to neat PLA and untreated fiber biocomposites.

In other work, tensile strength and Charpy notched strength were evaluated for PLA biocomposites with a variety of types of natural fiber: abaca fibers, man-made cellulose, jute and flax fibers. Authors observed that increasing the content of fibers up to 30 wt% the composite's stiffness significantly increases as well as tensile and impact strengths with respect to neat PLA [127]. The same improvement in mechanical properties was reported by Choie and Lee [128] using ramie fibers and PLA resin as matrix.

Tensile strength, Young's modulus and impact strength of short hemp fibre reinforced PLA biocomposites increased with increased fibre content (10–30 wt.%) as well as with the application of surface fiber treatments like alkali and silane treatments. It was found that PLA could be reinforced with a maximum of 30 wt.% fibres using conventional injection moulding, but could not be processed at higher fibre contents due to poor melt flow of the compounded materials [123]. In Table 5 the best results of each reference for some mechanical properties of PLA biocomposites with vegetable fiber are summarized.

As shown in Table 5, PLA biocomposites have shown different mechanical properties. Kenaf and hemp fiber PLA biocomposites showed a significantly increase in tensile strength and Young's modulus while a decrease in impact strength with respect to neat PLA was also reported [129]. In this work, neat PLA showed a tensile strength of 30.1 MPa, Young Modulus of 3.6 GPa and 24.4 kJ/m^2 for unnotched Charpy impact strength. The same observation was achieved by Oksman et al. [130] for unnotched Charpy impact strength of PLA biocomposite (12 kJ/m^2) with respect to neat PLA (15 kJ/m^2). Different values for neat PLA mechanical properties were reported and they depend mainly on inherent PLA properties (average molar mass, density, etc.) as well as the manufacturing process. Nevertheless, some authors have already observed an increase from a notched impact test for PLA biocomposites [82, 123, 124, 131] for different types of vegetable fibers.

Biodegradable composites have showed insufficient impact strength, preventing a broader field of application of these materials in automotive sector and in electronic devices. However, PLA reinforced with a man-made cellulose (Cordenka®) produced a biocomposite which have met performance requirements, especially for impact properties (72 kJ/m^2 for unnotched Charpy impact strength), that can be used in automotive and electronic industry [132]. Authors [129] also reported PLA biocomposites with man-made cellulose that have shown good tensile and impact properties and they can be used in different fields of application like household appliances and in bumpers in the automobile industry.

Fiber and Content (wt%)	Interface Treatment	Manufacturing Process	Tensile strength (MPa)	Young's modulus, (GPa)	Impact strength (kJ/m²)	Reference
Abaca (30)	Untreated fibers	Extrusion followed by injection molding	74.0	8.0	5.0 (notched Charpy)	124
Bamboo (20)	Plasma and silane coupling	a filmstacking procedure	90	1.8	-	71
Flax (30)	Enzime retting of fiber	Extrusion followed by compression molding	53	8.3	12 (unnotched Charpy)	130
Hemp (30)	Mercerized fiber	Extrusion followed by injection molding	75.5	8.2	2.64 (notched Charpy)	123
Hemp (40)	Untreated fibers	Roller carding with PLA followed by compression molding	57.5	8	9.5 (unnotched Charpy)	129
Jute (30)	Untreated fibers	Extrusion followed by injection molding	81.9	9.6	4.8 (notched Charpy)	124
Kenaf (40)	Untreated fibers	Roller carding with PLA followed by compression molding	52.9	7.1	9.0 (unnotched Charpy)	129
Kenaf (30)	5 wt% Coupling agent (maleic anhydride grafted PLA)	Internal mixing followed by compression molding	-	-	3.46 ± 0.13 (notched Charpy)	131
Man-made cellulose (Lyocell) (40)	Untreated fibers	Roller carding with PLA followed by compression molding	81.8	6.8	39.7 (unnotched Charpy)	129
Man-made cellulose (30)	Untreated fibers	Extrusion followed by injection molding	92	5.8	8.0 (notched Charpy)	124

Table 5. Tensile strength, Young's modulus and impact strength (room temperature) of PLA-based biocomposites with vegetable fibers.

Biocomposites that show high tensile strength and stiffness as well as low impact strength could be used in manufacture of furniture, boardings or holders for grinding discs and so on which are not subjected to high impact stress. Biocomposites that show the combination of properties as low tensile strength with high impact strength leads to application of these materials in interior parts in cars or safety helmets [129]. Also, kenaf fiber–reinforced PLA matrix biocomposites which the processing is based on injection molding have been used for spare tire covers and circuit boards [133] and these biocomposites were proposed to be used in an automotive headliner made from a 50/50 PLA/Kenaf fiber using a carding process [63].

The mechanical properties are thus among the most widely tested properties of natural fiber reinforced composites [2]. Compared with widespread research on mechanical properties of biocomposites, there are few reports on flame retardancy of biopolymers and biocomposites [134, 135]. The flame retardancy of ramie fiber reinforced PLA biocomposites was tested using halogen-free ammonium polyphosphate (APP). PLA biocomposites using flame-retardant treatment of ramie fibers have demonstrated a certain flame retardancy but cannot be classified by UL94 testing (Test for flammability of plastic materials for parts in devices and appliances) because of low APP loading (4.5 wt%). When PLA matrix is mixed in a extruded with APP, biocomposites with treated or non-treated ramie fibers and having the same APP loading (10.5 wt%) achieved V-0 rating (short burning time, no dripping; self-extinguishing). Low loading of APP does not adversely affect the mechanical performance of PLA/ramie biocomposites [136]. Other authors [137] also studied PLA biocomposites using plasma-treated coconnut fiber and prepared using the commingled yarn method. As expected, plasma-treated coconut fibers improved mechanical properties like tensile strength and modulus of biocomposites compared to neat PLA, but no significant changes on the fire retardant properties was achieved for the biocomposites with respect to neat PLA, according to the limiting oxygen index (LOI) value: around 25 for neat PLA and 10 wt% treated coconut fiber biocomposite. Generally, when the LOI value is greater than 26, materials can be considered to have flame retardancy [134].

Nanoreinforcements were also tested in fully biodegradable biocomposites of PLA matrix. These biocomposites help to provide new food packaging materials with improved mechanical, barrier, antioxidant and antimicrobial properties [138]. The addition of cellulose nanowhiskers to PLA matrix reduced the water permeability by up to 82% and the oxygen permeability by up to 90% with only 3 wt% of nanofiller content [139]. Moreover, the incorporation of organomodified mica-based clay to PLA matrix enhanced barrier properties to UV light; besides other barrier properties.This property is highly important for food packaging as protection against light which is a basic requirement to preserve the quality of many food products [140].

In previous research, PLA matrix was reinforced by 5wt% microcrystalline cellulose or 5wt% commercial organically modified bentonite (layered silicate) [141]. The bionanocomposite reinforced by bentonite showed great improvements in tensile modulus and strength as well as a decrease in oxygen permeability whereas the bionanocomposite reinforced with microcrystalline cellulose only showed a tendency to improve strength as well as a reduction in elongation at break. No changes for oxygen permeability were observed. This was attributed to the larger surface area of bentonite that allows interaction with a larger amount of PLA chains.

In other work, the presence of a surfactant favoured the dispersion of cellulose nanocrystals in the PLA matrix, yielding bionanocomposites with higher tensile modulus and strength. The addition of silver nanoparticles to the bionanocomposite did not enhance these mechanical properties. Besides, an antibacterial activity against *Staphylococcus aureus* and *Escherichia coli* cells was detected for ternary systems, indicating that these bionanocomposites have great potencial to be applied in food packaging when an antibacterial effect is required [95].

Polylactides and their copolymers were been widely reported to be used in the fields of orthopedic and reconstructive surgery due to its biodegradability and better features for use in the human body (nontoxicity) [142, 143]. According to Walker et al. [144], polylactides degrade in vivo by hydrolytic mechanisms of the ester bonds into lactic acid which is processed through metabolic pathways and is eliminated from the body through the renal and/or respiratory mechanisms. PLLA constructs have a longer degradation time when compared to other polymers, having shown to be present at 3 years after implantation. Its structural characteristics have proven useful for the construction of orthopedic hardware.

Bionanocomposites of hydroxyapatite (HPA) nanospheres which is the main inorganic constituent of natural bone and PLLA microspheres were tested for biomedical application to produce scaffolds using a laser sintering process [145]. HPA particles can reinforce polymer matrices and decrease the degradation rate of PLA [146]. Also, other work showed that PLA/organoclay bionanocomposites have enhanced their thermomechanical properties and gas barrier properties with respect to neat PLA and their biodegradation rate depends on the organoclay nature, organoclay content, organoclay dispersion as well as the organic modifier used to treat the nanoclay [147]. The relative hydrophilicity of the clay layers has been shown to play a key role in the hydrolytic degradation of the PLA chains [148].

Biodegradability of flax fiber reinforced PLA based biocomposites in presence of amphiphilic additives like benzilic acid, mandelic acid, dicumyl peroxide (DCP) and zein protein was investigated by soil burial test with farmland soil. Authors reported that neat PLA films degraded rapidly compared to natural fiber reinforced biocomposites. But, regarding the use of amphiphilic additives, the higher loss in weight is obtained for flax reinforced PLA biocomposites in the presence of mandelic acid. In the presence of DCP, the biodegradability of the biocomposites was comparatively delayed. Depending on the end-uses of the biocomposites, suitable amphiphilic additives can be used as triggers for inducing controlled biodegradation [149].

The aerobic biodegradation of biocomposites of PLA, thermoplastic starch (TPS) and a blend of 75 wt% of PLA and 25 wt% of TPS with short natural fiber (coir) with and without the addition of maleic anhydride (MA) coupling agent were investigated under controlled composting conditions. TPS showed higher biodegradation rates than PLA, probably due to the TPS domains preferentially attacked by microorganisms. Besides, authors ascertained that coir fibers probably have no influence in the biodegradation process due to the slight differences in carbon dioxide produced for neat polymers and their biocomposites with coir fiber. Also, the presence of coupling agent decreased the percentage of evolved CO_2 compared to biocomposites without coupling agent [150].

In other work, bacterial (*Burkholderia cepacia* bacteria) biodegradation studies were performed for biocomposites of PLA and mercerized banana fiber (BF) produced by melt blending followed by compression molding. Banana fibers were also treated with various silanes to improve their compatibility with PLA matrix. Authors reported improvements in tensile and impact strength of the biocomposites with respect to neat PLA. Weight loss experiments showed that PLA had 60% of degradation within a period of 25 days and all biocomposites showed higher degradation rates (80–100%). While biocomposites with untreated and alkaline-treated BF degraded almost completely, silane-treated biocomposites degraded at lower rates. Water absorption studies supported this evidence [151, 152].

4.2. PHBV biocomposites

Poly(hydroxyl-alkanoates) (PHAs).are a family of bacterial polyesters which poly(hydroxy-butyrate) (PHB) and its copolymer poly (3-hydroxybutyrate-co-3-valerate) (PHBV) make part. According to Bledzki and Jaszkiewicz [124], PHBV has been technologically developed to improve the known weaknesses of PHB like brittleness and poor processability.

Biocomposites of PHBV with wood and bamboo fibers were fabricated using extrusion followed by injection molding. Tensile and flexural modulus increased with fiber loading for biocomposites with the two kinds of fiber and no appreciable difference among the two fiber loadings (30 and 40 wt% fiber) was noticed. However, notch impact strength of PHBV decreased with the fiber addition and the reduction was greater in case of bamboo fiber biocomposites [153]. However, in other work biocomposites of PHBV and bamboo pulp fibers which were prepared by melt compounding and injection molding showed substantially increase of the impact strength by the addition of bamboo pulp fiber as well as increased tensile strength and modulus and flexural strength and modulus. The maleic anhydride grafted PHBV used as coupling agent improved polymer/fiber interactions and therefore resulted in increased strength and modulus. However, the toughness of the composites was substantially reduced due to the hindrance to fiber pullout [154]. Also, authors [124] reported an increase of the impact strength for PHBV biocomposites using 30 wt% of man-made cellulose, abaca and jute fibers at 23ºC and also at -30 ºC. The most pronounced results were obtained with man-made cellulose. PHBV was blended with 27.6 wt% of poly (butylene adipate-co-butylene terephtalate) (PBAT) and 2.4 wt% of processing aids. Moreover, tensile strength and modulus were increased.

In recent work, PHBV was blended with PBAT using extrusion (in a twin-screw extruder) followed by injection molding. Biocomposites were performed with 20–40wt% switchgrass and the compatibilizer pMDI. With the addition of 25wt% switchgrass the tensile and flexural strengths of the biocomposite have improved. On increasing the fiber content to 30wt% and further to 40wt%, both tensile and flexural strength dropped but the modulus of the composites increased progressively with increasing fiber content. With regard to uncompatibilized composites, impact strength of 53 J/m was achieved for composites with 25wt% switchgrass because of the proper wetting achieved between the fiber and the matrix. Impact strength reduced with increase in fiber content. The use of the pMDI compatibilizerer in biocomposites with 30 wt% switchgrass promoted interfacial interactions between the matrix and the fiber and significantly improved the mechanical properties of the biocomposites. The addition of pMDI significantly increased the impact strength of the composites. The notched impact strength increased 80% compared to the uncompatibilized composite owing to the enhanced interfacial adhesion [155]. Also, by incorporation of biomass fiber reinforcement like corn straw, soy stalk and wheat straw into the PHBV by melt mixing technique, authors showed that the alkali treatment of wheat straw fibers enhanced strain, break and impact strength of PHBV composites by 35%, hardly increasing strength and modulus compared to their untreated counterparts. Authors also showed that the tensile and storage modulus of PHBV were improved by maximum 256% and 308%, respectively, with 30 wt% of the biomass and these values were much higher than the corresponding polypropylene (PP) composites [156].

Nanoparticles also have already been incorporated into PHBV matrix. Well-dispersed cellulose nanocrystals into PHBV matrix were obtained with simultaneous enhancements on the mechanical property and thermal stability of PHBV. Compared to neat PHBV, a 149% improvement in tensile strength and 250% increase in Young's modulus were obtained for the resulting nanocomposites with 10 wt% of cellulose nanocrystals [157]. Lower concentrations of cellulose nanowhiskers (0–4.6 wt%) were used to prepare PHBV bionanocomposites by solution casting [158]. The mechanical properties of the films increased with increasing cellulose nanowhiskers content until the content reached 2.3 wt %. Real permittivity of the composites also peaked at 2.3 wt % cellulose nanowhiskers over a wide spectrum of frequencies (0.01–10^6 Hz). These property transitions at 2.3% cellulose nanowhiskers content were due to the transition of cellulose nanowhiskers dispersion from homogeneous dispersion to agglomeration. Nevertheless, rheological results of the bionanocomposites indicated a transition point lower than 2.3% due to the formation of a biopolymer-fiber network in the composite melt.

Some authors [159] showed that the incorporation of low concentrations of nanoclays (5 wt%) and cellulose nanowhiskers (3 wt%) into PHBV matrix and other biodegradable matrices like PLA and polycaprolactone (PCL) resulted in improvements in oxygen permeability that can be very useful for food packaging. With respect to water permeability, authors showed that PHBV films with 1 wt% alpha cellulose fiber content had a water permeability drop of 71% compared to the unfilled material, whereas PHBV films with a fiber content of 10 wt% showed a water permeability reduction of around 52% due to fiber agglomeration. However, the lowest water and limonene permeability coefficient values were obtained for the bionanocomposites containing 5 wt% of clay due to the good morphology for these nanocomposites. The same work also reported that mica-based nanoclays exerted certain UV/visible light blocking action in PLA and PHBV matrices. The blocking effect of PHBV in the UV-Vis region was higher than that of PLA since PHBV is a translucent material. Moreover, greater reductions in vapour permeability were attained for PHBV bionanocomposites with clay contents of 1 wt% [94]. Furthermore, the PHBV processing behavior could be improved with addition of montmorillonite nanoclay since the processing temperature range enlarged by lowering melting temperature with the increasing clay content. The tensile properties of the corresponding materials were improved by incorporation of 3wt% of clay [160].

Thus, in general many properties have been improved with the incorporation of fibers and mainly nanofibers and nanoclays into PHBV which are helpful to overcome many obstacles and enhance the efficiency in a diverse number of applications. In this way, it is found that nanofibers can induce fast regeneration of many tissues/organs in medical applications and improve the efficiency of many chemical and electronic applications [161].

PHA's family was related to be used in numerous biomedical applications, such as sutures, cardiovascular patches, wound dressings, scaffolds in tissue engineering, tissue repair/ regeneration devices, drug carriers and so on, but much deep studies [162]. PHBV bionano-composites were manufactured with various calcium phosphate-reinforcing phases for bone tissue regeneration while inducing a minimal inflammatory response. Authors showed that the addition of a mineral nano-sized reinforcing phase to PHBV reduced the proinflammatory response and also improved osteogenic properties with respect to pure PHBV [163].

With respect to biodegradation behaviour, biocomposites of PHBV matrix and 10, 20 and 25 wt% of peach palm particles were investigated [164]. Soil biodegradation tests were carried out according to ASTM G160-98 with test exposures of up to 5 months. The addition of peach palm particles reduced the maximum strength but improved the Young's modulus and also soil biodegradation tests indicated that the biocomposites degraded faster than the neat polymer due to the presence of cavities that resulted from introduction of the peach palm particles and that degradation increased with increasing particles content. These voids allowed for enhanced water adsorption and greater internal access to the soil-borne degrader micro-organisms. Similarly, other authors found that biocomposites with PHBV and wood fiber have higher degradation rates than the neat polymer [165]. On the other hand, some authors reported no significant difference between the degradability of PHBV and its composite with wheat straw using either Sturm tests or soil burial tests [166].

5. Conclusion

Due to the high demand for environmental sustainable products, researchers continue to seek materials derived from renewable resources that can be applied in a wide range of applications. This overview provided a survey of some of the current researches on the biocomposites area. Within this context, this chapter showed that there have been many attempts to produce biocomposites using natural reinforcements and biobased polymers since improvements in their mechanical, barrier and other properties can be accomplished through the use of reinforcement agents like vegetable fibers and nanoparticles (cellulose nanofiber or nano-clays). Vegetable fibers are generally submitted to chemical treatments, mostly alkaline and acid treatments in order to favour interfacial adhesion between polymer matrices and the fiber. Also, the use of coupling agents enhance adhesion by surface modification as well as they can produce grafting reactions between matrix and fiber. Moreover, the presence of polar groups in most biobased poymers contributes to better affinity to cellulosic groups of vegetable fibers. All these issues dramatically influence the mechanical properties of the biocomposites. With respect to nanoreinforcements, cellulose nanofibers and organic functionalized clays (orga-noclays) are the most used as fillers in bionanocomposites.

PLA based biocomposites are one of the most studied biocomposites and some researches showed that the use of vegetable fiber can improve the impact strengh of the PLA matrix, but insufficient strength values were found to enable their application in automotive sector and in electronic devices. PLA biocomposites with a man-made cellulose fiber that fulfill the requirements for mechanical properties were already reported and their use can be extended to diferent fields of application. The use of nanoreinforcements in PLA matrices produced bionanocomposites with remarkable mechanical, thermal, barrier, antioxidant and antimicrobial properties, presenting a new material with potential for food packaging application. The biodegradability of PLA biocomposites with vegetable fibers showed to be sensitive to the additives used in biocomposite processing. The presence of coupling agents provides lower degradation times than neat PLA. Also, depending on the nature of the amphilic additives, they may speed up or delay the biodegradation process. Researches with organoclay in bionanocomposites showed that their biodegradation rate depends on the nature, the content and the dispersion level of organoclay in the bionanocomposite as well as the nature of organic modifier of the clay.

PHBV based biocomposites also showed an increase in mechanical properties in presence of treated vegetable fibers and coupling agents. However, the incorporation of cellulose nanofibers and organoclays in PHBV matrix promoted greater improvements not only in mechanical properties but also in oxygen and water permeability. The bionanocomposites produced can be used in medical applications due to the faster regeneration of many tissues/organs and in many chemical and electronic applications. The specific use of organoclays also produced UV-Vis blocking effects and greater reductions in vapour permeability as well as processing behaviour improvements. The biodegradability of these bionanocomposites showed to be similar or faster than the neat PHBV matrix.

Therefore, bionanocomposites arised as a promising area that can overcome some of the drawbacks of biobased polymers and their biocomposites since the use of nanoparticles generally promotes greater improvements in many properties with respect to biocomposites. However developments must be performed on processing techniques and key research callenges like nanoparticles dispersion into biopolymers. Thus, the construction of a biocomposite/bionanocomposite is not a simple process and it needs the knowledge of the real contribution of each composite phase for property tuning. Moreover, biocomposites/bionanocomposites will be only attractive if material and process costs are competitive compared to conventional composites which use petrochemical resources.

Acknowledgements

The author Derval dos Santos Rosa thanks FAPESP – Process no 2012/13445-8 and UFABC for support.

Author details

Derval dos Santos Rosa[1] and Denise Maria Lenz[2*]

1 Universidade Federal do ABC, SP, Brazil

2 Universidade Luterana do Brasil, RS, Brazil

References

[1] Chand, N and Fahim, M Tribology of natural fiber polymer composites. Chapter 1, 1-58, Woodhead Publishing and CRC Press (2008).

[2] Faruk, O, Andrzej, K, Bledzki, H-P, Fink, M S Biocomposites reinforced with natural fibers: 2000–2010. Progress in Polymer Science, 37(11)1552-1596 (2012).

[3] Kalia, S, Dufresne, A, Cherian, B M, Kaith, B S, Avérous, L, Njuguna, J, Nassiopoulos, E Cellulose-Based Bio- and nanocomposites: a review. International Journal of Polymer Science (2011) Article ID 837875, 35 pages doi:10.1155/2011/837875.

[4] John, M J, Thomas, S Biofibres and biocomposites. Carbohydrate Polymers, 71: 343-364 (2008).

[5] Fowler, P A, Hughes, J M, Elias, R M Review Biocomposites: technology, environmental credentials and market forces. Journal of the Science of Food and Agriculture, 86(12):1781-1789 (2006).

[6] Bledzki, A K, Gassan, J Composites reinforced with cellulose-based fibres, Progress in Polymer Science, 24(2): 221-274 (1999).

[7] Ruiz-Hitzky, E, Darder, M, Aranda, P Progress in bionanocomposite materials. In: Annual Review Nano Research Volume 3, Chapter 3, 149-189, Guozhong Cao, Qifeng Zhang and C. Jeffrey Brinker(Ed) World Scientific Publishing, Singapore (2010).

[8] Malviya, R, Srivastava, P, Bansal, V and Sharma, P K Formulation, evaluation and comparison of sustained release matrix tablets of diclofenac sodium using natural polymers as release modifier, International Journal of Pharma and Biosciences,1(2): 1-8 (2010).

[9] Malviya, R, Srivastava, P, Bansal M and Sharma P K Preparation and evaluation of disintegrating properties of Cucurbita maxima pulp powder. International J. Pharmaceutical Sci., 2(1):395-399 (2010).

[10] Šimkovic, I What could be greener than composites made from polysaccharides? Carbohydrate Polymers, 74 (4): 759-762 (2008).

[11] Rosa, D S, Guedes, C G F, Casarin, F, Braganca, F C The effect of the Mw of PEG in PCL/CA blends. Polymer Testing, 24(5): 542–548 (2005).

[12] Klemm, D , Heublein, B, Fink, H-P, Bohn, A Cellulose: fascinating biopolymer and sustainable raw material Angewandte Chemie International Edition, 44 (22): 3358-3393 (2005).

[13] Edgar, K J , Buchanan, C M, Debenham, J S, Rundquist, P A, Seiler, B D, Shelton, M C, Tindall, D Advances in cellulose ester performance and application Progress in Polymer Science, 26 (9):1605-1688 (2001).

[14] Ignácio C and Barros D M Membranas celulósicas – efeitos da concentração de polí-mero-solvente-não solvente na morfologia da membrana. 8º Brazillian Polymer Congress, Águas de Lindóia, 624-625 (2005).

[15] Mulinari, D R, Silva M L C P, Silva G L J P Preparação e caracterização dos compósi-tos celulose branqueada/ZrO2nH2O preparados pelos métodos da precipitação con-vencional e precipitação em solução homogênea. 8º Brazillian Polymer Congress, Águas de Lindóia, 245-246 (2005).

[16] Vroman, I and Tighzert, L Biodegradable Polymers, Materials, 2: 307-344 (2009) doi: 10.3390/ma2020307.

[17] Rosa D S, Carvalho C L, Gaboardi, M L F, Rezende, M I, Tavares B, Petro M S M, Calil M R Evaluation of enzymatic degradation based on the quantification of glu-cose in thermoplastic starch and its characterization by mechanical and morphologi-cal properties and NMR measurements. Polymer Testing, 27(7): 827-834 (2008).

[18] Rosa, D S, Guedes, C G F, Carvalho C L Processing and thermal, mechanical and morphological characterization of post-consumer polyolefins/thermoplastic starch blends. Journal of Materials Science, 42(2): 551–557 (2007).

[19] Pedroso, A G and Rosa, D S Mechanical, thermal and morphological characterization of recycled LDPE/corn starch blends. Carbohydrate Polymers, 59 (1): 1-9 (2005).

[20] Rosa, D S, Guedes, C G F, Volponi, J E Biodegradation and dynamic mechanical properties of starch gelatinization in poly(-caprolactone)/corn starch blends. Journal of Applied Polymer Science, 102 (1): 825-832 (2006).

[21] Rosa, D S, Guedes C G F, Carvalho, C L Processing and thermal, mechanical and morphological characterization of post-consumer polyolefins/thermoplastic starch blends. Journal of Materials Science, 42(2): 551–557 (2007).

[22] Furusaki E., Ueno Y., Sakairi N., Nishi N. and Tokura, S. Facile Preparation and in-clusion ability of a chitosan derivative bearing carboxymethyl-b-cyclodextrin. Carbo-hydrate Polymers, 9: 29-34 (1996),

[23] Hirano, S and Nagano, N Effects of chitosan, pectic acid, lysozyme and chitinase on the growth of several phytopathogens. Agricultural and Biological Chemistry, 53: 3065-3066 (1989).

[24] Kumar, M N V R A review of chitin and chitosan applications. Reactive & Functional Polymers, 46 (1): 1-27 (2000).

[25] Piermaria, J A, Pinotti, A, Garcia, M A, Abraham, A G Films based on kefiran, an exopolysaccharide obtained from kefir grain: development and characterization. Food Hydrocoll., 23: 684-690 (2009).

[26] Chandra, R, Rustgi, R Biodegradable polymers. Progress in Polymer Science, 23: 1273-1335 (1998).

[27] Andrade, C T, Lopes, L Polímeros de origem microbiana: polissacarídeos bacterianos. Revista de Química Industrial, 703:19-23 (1995).

[28] Serafim S L, Lemos P C, Reis, M A M Produção de bioplásticos por culturas microbianas mistas. Boletim de Biotecnologia, 76:15-20 (2003).

[29] Pradella, J G C Biopolímeros e intermediários químicos. Technical Report nº 84396-205, Centro de Tecnologia de Processos e Produtos, Laboratório de Biotecnologia Industrial – LBI/CTPP (2006).

[30] Coutinho, B C, Miranda, G B, Sampaio, G R, Souza, L B S, Santana, W J, Coutinho, H D M A importância e as vantagens do polihidroxibutirato (plástico biodegradável). Holos 3: 76-81 (2004).

[31] Lenz, R W, Marchessault, R H Bacterial polyesters: biosynthesis, biodegradable plastics and biotechnology. Biomacromolecules 6 (1):1-8 (2005).

[32] Rosa, D S, Penteado, D F, Calil, M R Propriedades térmicas e biodegradabilidade de PCL e PHB em um pool de fungos. Revista Ciência & Tecnologia, 15: 75-80 (2000).

[33] Corrêa, M C S, Rezende, M L, Rosa, D S, Agnelli, J A M, Nascente, P A P Surface composition and morphology of poly(3-hydroxybutyrate) exposed to biodegradation. Polymer Testing 27 (4): 447-452, 2008.

[34] Quental, A C, de Carvalho, F P, Rezende, M I, Rosa, D S, Felisberti M I Aromatic/aliphatic polyester blends. Journal of Polymers and the Environment, 18 (3): 308-317 (2010).

[35] Luckachan, G. E., Pillai C. K. S. Biodegradable polymers- a review on recent trends and emerging perspectives Journal of Polymers and the Environment , 19: 637–676 (2011).

[36] Keshavarz, T and Roy, I Polyhydroxyalkanoates: bioplastics with a green agenda Current Opinion in Microbiology, 13:321-326 (2010).

[37] Rosa, D S and Pântano Filho, R Biodegradação: um ensaio com polímeros. Moara (Ed.), Itatiba, S.P; Editora Universitária São Francisco (Ed.), Bragança Paulista, S.P. (2003).

[38] Lotto, N T Dimensionamento da degradação dos polímeros PHB e PHB-V através da variação da rugosidade. (Master Dissertation), 78 p. Universidade São Francisco, S. P., Brazil (2003).

[39] Rosa, D S, Franco, B L M, Calil, M R Biodegradabilidade e propriedades mecânicas de novas misturas poliméricas. Polímeros: Ciência e Tecnologia, 2 (11): 82-88 (2001).

[40] Lotto, N T, Calil, M R, Guedes. C G F, Rosa, D S The effect of temperature on the biodegradation test. Materials Science and Engineering: C, 24(5): 659-662 (2004).

[41] Vogelsanger, N, Formolo, M C, Pezzin, A P T, Schneider, A L S, Furlan, A A, Bernardo, H P, Pezzin S H, Pires, A T N, Duek, E A R Blendas biodegradáveis de poli(3-hidroxibutirato)/poli(-caprolactona): obtenção e estudo da miscibilidade. Materials Research, 6 (3):359-365 (2003).

[42] Raghavan, A D Characterization of Biodegradable Plastics. Polymer-Plastics Technology and Engineering, 34(1): 41-63 (1995).

[43] Zenkiewicz, M, Richert, J, Różański, A Effect of blow moulding ratio on barrier properties of polylactide nanocomposite films. Polymer Testing, 29(2): 251-257 (2010).

[44] Bhatia, A, Gupta, R, Bhattacharya, S, Choi, H. Effect of clay on thermal, mechanical and gas barrier properties of biodegradable poly(lactic acid)/poly(butylene succinate) (PLA/PBS) nanocomposites. International Polymer Processing, 25 (1):5-14 (2010).

[45] Martino, V P, Ruseckaite, R A, Jiménez, A, Averous, L Correlation between composition, structure and properties of poly(lactic acid)/polyadipate-based nano-Biocomposites. Macromolecular Materials and Engineering, 295(6): 551-558 (2010).

[46] Rasala, R M, Janorkarc, A V, Hirta, D E Poly(lactic acid) modifications Progress in Polymer Science, 35:338–356 (2010).

[47] Henry, F, Costa, L C, Devassine, M The evolution of poly(lactic acid) degradability by dielectric spectroscopy measurements European Polymer Journal, 41(9): 2122–2126 (2005).

[48] Ulery, B D, Nair, L S, Laurencin, C T Biomedical applications of biodegradable polymers Journal of Polymer Science Part B: Polymer Physics, 49: 832–864 (2011).

[49] Bragança, F C and Rosa, D S Thermal, mechanical and morphological analysis of poly(3-caprolactone), cellulose acetate and their blends, Polymers for Advances Technology 14(10):669–675 (2003).

[50] Landry, C J T, Lum, K K, O'Reilly, J M Physical aging of blends of cellulose acetate polymers with dyes and plasticizers, Polymer, 42 (13):5781-5792 (2001).

[51] Wang, Z, Qu, B, Fan, W, Hu, Y, Shen, X Effects of PE-g-DBM as a compatibilizer on mechanical properties and crystallization behaviors of magnesium hydroxide- based LLDPE blends. Polymer Degradation and Stability, 76(1): 123-128 (2002).

[52] Kim, C-H, Cho, K Y, Park, .J-K Grafting of glycidil methacrylate onto polycaprolactone: preparation and characterization. Polymer, 42(12): 5135-5142 (2001).

[53] Liu, W, Wang, Y L, Sun, Z Effects of polyethylene-grafted maleic anhydride (PE-g-MA) on thermal properties, morphology, and tensile properties of low-density polyethylene (LDPE) and corn starch blends, Journal of Applied Polymer Science, 88 (13): 2904–2911 (2003),.

[54] Mucci, V, Pérez, J, Vallo, C I Preparation and characterization of light-cured methacrylate/montmorillonite nanocomposites Polymer International, 60 (2): 247-254 (2011).

[55] Sreekumar, P A and Thomas, S Matrices for natural-fibre reinforced composites. In: Properties and performance of natural-fibre composites. Chapter 2, 67-126, Kim L. Pickering (Ed) Woodhead Publishing Limited and CRC Press (2008).

[56] Kozlowski, R and Wladyka-Przybylak, M Uses of natural fiber reinforced plastics. In: Natural Fibers, Plastics and Composites. Chapter 14, 249-271, Wallenberger FT, Weston NE (Ed). Kluwer Academic Publishers: Dordrecht (2004).

[57] Yang, L, Hu, Y, Lu, H, Song, L Morphology, thermal, and mechanical properties of flame-retardant silicone rubber/montmorillonite nanocomposites. Journal of Applied Polymer Science, 99(6): 3275–3280 (2006).

[58] Duhovic, M, Peterson, S, Jayaraman, K Natural-fibre-biodegradable polymer composites for packaging In: Properties and performance of natural-fibre composites. Chapter 9, 301-329, Kim L. Pickering (Ed), Woodhead Publishing Limited and CRC Press (2008).

[59] Satyanarayana, K G, Arizaga, G G C, Wypych, F Biodegradable composites based on lignocellulosic fibers- an overview. Progress in Polymer Science, 34(9): 982-1021 (2009),.

[60] Spinacé, M A S, Lambert, C S, Fermoselli, K K G, De Paoli, M-A. Characterization of lignocellulosic curaua fibres. Carbohydrate Polymers, 77(1): 47-53 (2009).

[61]

[62] Sánchez, C Lignocellulosic residues: Biodegradation and bioconversion by fungi. Biotechnology Advances, 27(2):185-194 (2009).

[63] Pilla, S Engineering applications of bioplastics and biocomposites – an overview. In: Handbook of Bioplastics and Biocomposites - Engineering Applications, chapter 1, 1-14, Srikanth Pilla (Ed), Scrivener Publishing LCC and John Wiley & Sons, Massachusetts and New Jersey (2011).

[64] Lee, B-H, Kim, H-S, Lee, S, Kim, H-J, Dorgan, J R Bio-composites of kenaf fibers in polylactide: Role of improved interfacial adhesion in the carding process. Composites Science and Technology, 69(15-16): 2573–2579 (2009).

[65] Stelte, W, Sanadi, R A Preparation and characterization of cellulose nanofibers from two commercial hardwood and softwood pulps. Industrial and Engineering Chemical Research, 48 (24):11211–11219 (2009).

[66] Hearle, J W S and Sparrow, J T Mechanics of the extension of cotton fibers. I. Experimental studies of the effect of convolutions. Journal of Applied Polymer Science, 24 (6): 1465-1477 (1979).

[67] Silva, R V and Aquino, E M F Curaua Fiber: a new alternative to polymeric composites Journal of Reinforced Plastics and Composites, 27(1): 103-112 (2008).

[68] Santos, P A, Spinacé, M A S, Fermoselli, K K G, De Paoli, M.-A Polyamide-6/vegetable fiber composite prepared by extrusion and injection molding. Composites Part A: Applied Science and Manufacturing , 38 (12): 2404–2411 (2007).

[69] Bismark, A, Mishra, S, Lampke,, T. Plant Fibers as reinforcement for green composites. In: Natural fibers, biopolymers and biocomposites, Chapter 2, 37-108, A. K. Mohanty, M. Misra and L. T. Drzal (Eds) Taylor & Francis, CRC Press (2005).

[70] Li. X, Tabil, L G, Panigrahi, A S Chemical Treatments of Natural Fiber for Use in Natural Fiber-Reinforced Composites: A Review. Journal of Polymer and the Environment,15 (1): 25–33 (2007).

[71] Kalia, S, Kaith, B S, Kaur, I. Pretreatments of natural fibers and their application as reinforcing material in polymer composites—a review. Polymer Engineering & Science, 49 (7): 1253–1272 (2009).

[72] Ma, H and Joo, C W Influence of surface treatments on structural and mechanical properties of bamboo fiber-reinforced poly(lactic acid) biocomposites. Journal of Composite Materials, 45(23): 2455-2463 (2011).

[73] Franco, P J H, Valadez-González, A Fiber-matrix adhesion in natural fiber composites In: Natural fibers, biopolymers and biocomposites, Chapter 6, 177-230, A. K. Mohanty, M. Misra and L. T. Drzal (Eds) Taylor & Francis, CRC Press (2005).

[74] Singha, A S and Rana, A K Effect of silane treatment on physicochemical properties of lignocellulosic C. indica fiber Journal of Applied Polymer Science, 124(3): 2473–2484 (2012).

[75] Campos, A, Marconcini, J M, Martins-Franchetti, S M, Mattoso, L H C The influence of UV-C irradiation on the properties of thermoplastic starch and polycaprolactone biocomposite with sisal bleached fibers. Polymer Degradation and Stability, 97 (10): 1948-1955 (2012).

[76] Stocchi, A, Bernal, C, Vazquez, A, Biagotti, J, Kenny, J A silicone treatment compared to traditional natural fiber treatments: effect on the mechanical and viscoelastic prop-

erties of jute–vinylester laminates. Journal of Composite Materials, 41 (16): 2005-2024 (2007).

[77] Moraes, A G O, Sierakowski, M R, Amico, S C The novel use of sodium borohydride as a protective agent for the chemical treatment of vegetable fibers Fibers and Polymers. 13 (5): 641-646 (2012) DOI: 10.1007/s12221-012-0641-7.

[78] Mukherjee, A, Ganguly P K, Sur, D J Structural mechanics of jute: the effects of hemicellulose or lignin removal. Journal of the Textile Institute. 84 (3):348-353 (1993).

[79] Rong, M Z, Zhang, M Q, Liu, Y., Yang, G C, Zeng, H M The effect of fiber treatment on the mechanical properties of unidirectional sisal-reinforced epoxy composites. Composites Science and Technology, 61 (10): 1437–1447 (2001).

[80] Kim, H-S, Lee, B-H, Choi, S-W, Kim, S, Kim, H-J The effect of types of maleic anhydride-grafted polypropylene (MAPP) on the interfacial adhesion properties of bioflour-filled polypropylene composites. Composites: Part A, 38 (6): 1473–1482 (2007).

[81] Chang, S Y, Ismail, H, Ashan, Q Effect of maleic anhydride on kenaf dust filled polycaprolactone/ thermoplastic sago starch composites. Bioressurces, 7(2):1594-1616 (2012).

[82] Xie, Y, Hill, C A S, Xiao, Z, Militz, H, Carsten, M Silane coupling agents used for natural fiber/polymer composites: A review. Composites: Part A 41(7): 806–819 (2010),.

[83] Huda, M S, Drzal, L T, Mohanty, A K, Misra, M Effect of fiber surface-treatments on the properties of laminated biocomposites from poly(lactic acid) (PLA) and kenaf fibers. Composites Science and Technology 68 (2): 424–432 (2008)..

[84] Chen, D, Li, J, Ren, J Influence of fiber surface-treatment on interfacial property of poly(l-lactic acid)/ramie fabric biocomposites under UV-irradiation hydrothermal aging. Materials Chemistry and Physics, 126 (3): 524–531 (2011).

[85] Jayamol, G, Sreekala, M S, Thomas, S A review on interface modification and characterization of natural fiber reinforced plastic composites. Polymer Engineering & Science 41 (9): 1471–1485 (2001)..

[86] Yu, L, Dean, K, Li, L Polymer blends and composites from renewable resources. Progress in Polymer Science 31(6): 576–602 (2006).

[87] Avérous, L Biocomposites based on biodegradable thermoplastic polyester and lignocellulose fibers In: Cellulose Fibers: Bio- and Nano-Polymer Composites: Green Chemistry and Technology. Chapter 17,453- 478, Susheel Kalia, B. S. Kaith, Inderjeet Kaur(Eds), Springer, Heidelberg (2011).

[88] Harnnecker, F, Rosa, D S, Lenz, D M Biodegradable Polyester-Based Blend Reinforced with Curaua Fiber: Thermal, Mechanical and Biodegradation Behaviour Journal of Polymers and the Environment, 20(1): 237-244 (2012).

[89] Eichhorn, S J, Dufresne, A, Aranguren, M, Marcovich, N E, Capadona, J R, Rowan, S J, Weder, C, Thielemans, W, Roman, M, Renneckar, S, Gindl, W, Veigel, S, Keckes, J,

Yano, H, Abe, K, Nogi, M, Nakagaito, A N, Mangalam, A, Simonsen, J, Benight, A S, Bismarck, A, Berglund, L A, Peijs, T Review: current international research into cellulose nanofibres and nanocomposites Journal of Materials Science, 45(1):1–33 (2010).

[90] Szczęsna-Antczak, M, Kazimierczak, J, Antczak, T Nanotechnology - Methods of Manufacturing Cellulose Nanofibres. Fibers & Textiles in Eastern Europe, 20, 2(91): 8-12 (2012).

[91] Iguchi, M, Yamanaka S, Budhiono A Bacterial cellulose - a masterpiece of nature's arts. Journal of Materials Science, 35(2):261-270 (2000).

[92] Rusli R., Shanmuganathan K., Rowan S. J., Weder C, Eichhorn S J. Stress Transfer in Cellulose Nanowhisker Composites - Influence of Whisker Aspect Ratio and Surface Charge. Nature 472: 334-337 (2011).

[93] European Commission, Nanotechnology Research needs on nanoparticles. Proceedings of the workshop held in Brussels, 25-26.01.2005.

[94] Halley, P J and Dorgan, J R Next-generation biopolymers: Advanced functionality and improved sustainability MRS Bulletin, 36: 687-691 (2011).

[95] Masoodi, R, Hajjar, R E, Pillai, K M, Sabo, R Mechanical characterization of cellulose nanofiber and bio-based epoxy composite. Materials and Design, 36: 570-576 (2011).

[96] Fortunati, E, Armentano, I, Zhou, Q, Iannoni, A, Saino, E, Visai, L, Berglund, L A, Kenny, J M Multifunctional bionanocomposite films of poly(lactic acid), cellulose nanocrystals and silver nanoparticles. Carbohydrate Polymers, 87(2): 1596–1605 (2012).

[97] Cherian, B M, Leão, A L, de Souza, S F, Costa, L M M, Olyveira, G M, Kottaisamy, M, Nagarajan, E R, Thomas, S Cellulose nanocomposites with nanofibers isolated from pineapple leaf fibers for medical applications Carbohydrate Polymers, 86(4): 1790–1798 (2011).

[98] Souza Lima, M M and Borsali, R. Rodlike Cellulose Microcrystals: Structure, Properties, and Applications. Macromolecular Rapid Communications, 25 (7): 771-787 (2004).

[99] Habibi, Y, Lucia, L A, Rojas, O J Cellulose Nanocrystals: Chemistry, Self-Assembly, and Applications Chemical Reviews, 110: 3479–3500 (2010).

[100] George, J, Ramana, K V, Bawa, A S and Siddaramaiah. Bacterial cellulose nanocrystals exhibiting high thermal stability and their polymer nanocomposites. International Journal of Biological Macromolecules, 48(1): 50-57 (2011).

[101] Jonoobi, M, Harun, J, Mathew, A P, Oksman, K Mechanical properties of cellulose nanofiber (CNF) reinforced polylactic acid (PLA) prepared by twin screw extrusion, Composites Science and Technology, 70(12): 1742-1747 (2010).

[102] Siró, I and Plackett, D Microfibrillated cellulose and new nanocomposite materials: a review. Cellulose, 17 (3): 459–494 (2010).

[103] Oksman, K, Mathew, A P, Sain, M Novel bionanocomposites: processing, properties and potential applications. Plastics, Rubber and Composites, 38(9-10): 396-405 (2009).

[104] Das, K, Ray, D, Banerjee, C, Bandyopadhyay, N R, Sahoo, S, Mohanty, A K, Misra, M Physicomechanical and Thermal Properties of Jute-Nanofiber-Reinforced Biocopolyester Composites. Industrial & Engineering Chemical Research, 49(6): 2775–2782 (2010).

[105] Wang, T and Drzal, L T Cellulose-nanofiber-reinforced poly(lactic acid) composites prepared by a water-based approach. Applied Materials & Interfaces, 4(10): 5079–5085 (2012).

[106] Okubo, K, Fujii, T, Thostenson, E T Multi-scale hybrid biocomposite: processing and mechanical characterization of bamboo fiber reinforced PLA with microfibrillated cellulose Composites Part A, 40(4): 469-475 (2009).

[107] Cherian, B M, Leão, A L, de Souza, S F, Thomas, S, Pothan, L A, Kottaisamy, M Isolation of nanocellulose from pineapple leaf fibers by steam explosion. Carbohydrate Polymers, 81(3): 720–725, (2010).

[108] Blaker, J J, Lee, K-Y, Bismarck, A. Hierarchical Composites Made Entirely from Renewable Resources Journal of Biobased. Materials and Bioenergy, 5(1): 1-16 (2011).

[109] Sorrentino, A, Gorrasi, G, Vittoria, V Potential Perspectives of Bio-Nanocomposites for Food Packaging Applications, Trends in Food Science & Technology, 18(2): 84-95 (2007).

[110] Chrissafis, K and Bikiaris, D Can nanoparticles really enhance thermal stability of polymers? Part I: An overview on thermal decomposition of addition polymers. Thermochimica Acta, 523(1): 1–24 (2011).

[111] Choy, J H, Choi, S J, Oh, J M, Park, T Clay minerals and layered double hydroxides for novel biological applications. Applied Clay Science, 35(1-3): 122-132 (2007).

[112] Liu, Z and Erhan, S Z "Green" composites and nanocomposites from soybean oil. Materials Science and Engineering A, 483–484: 708–711 (2008).

[113] Annabi-Bergaya, F Layered clay minerals. Basic research and innovative composite applications. Microporous and Mesoporous Materials, 107(1-2): 141–148 (2008).

[114] Fu, H-K, Huang. C-F, Huang, J-M, Chang, F-C Studies on thermal properties of PS nanocomposites for the effect of intercalated agent with side groups. Polymer, 49(5): 1305-1311 (2008).

[115] Rodriguez, F, Sepulveda, H M, Bruna, J, Guarda, A, Galotto, M J Development of Cellulose Eco-nanocomposites with Antimicrobial Properties Oriented for Food

Packaging. Packaging Technology and Science, 2012 DOI: 10.1002/pts.1980. http://onlinelibrary.wiley.com/doi/10.1002/pts.1980/full (acessed 20 July 2012).

[116] Arora, A and Padua, G W Review: Nanocomposites in Food Packaging Journal of Food Science, 75(1): R43–R49 (2010).

[117] Rajkiran, R, Tiwari, K C, Khilar, U N Synthesis and characterization of novel organomontmorillonites. Applied Clay Science, 38(3–4): 203–208 (2008).

[118] Kim, J-T, Lee, D-Y, Oh, T-S, Lee, D-H Characteristics of nitrile–butadiene rubber layered silicate nanocomposites with silane coupling agent Journal of Applied Polymer Science, 89(10): 2633–2640 (2003).

[119] Panagiotis, I, Xidas, K, Triantafyllidis, S Effect of the type of alkylammonium ion clay modifier on the structure and thermal/mechanical properties of glassy and rubbery epoxy–clay nanocomposites European Polymer Journal, 46(3): 404–417 (2010).

[120] Piscitelli, F, Scamardella, A M, Valentina, R, Lavorgna, M, Barra, G, Amendola, E Epoxy composites based on amino-silylated MMT: The role of interfaces and clay morphology Journal of Applied Polymer Science, 124(1): 616–628 (2012).

[121] Tjong, S C Structural and mechanical properties of polymer nanocomposites, Materials Science and Engineering: R: Reports, 53(3–4): 73–197 (2006).

[122] Bordes, P, Pollet, E, Avérous, L Nano-biocomposites: Biodegradable polyester/nanoclay systems Progress in Polymer Science, 34(2): 125–155 (2009)

[123] Kumar, K A A, Sreekala, M S, Arun, S Studies on Properties of Bio-Composites from Ecoflex/Ramie Fabric-Mechanical and Barrier Properties Journal of Biomaterials and Nanobiotechnology, 3: 396-404 (2012).

[124] Sawpan, M A, Pickering, K L, Fernyhough, A Improvement of mechanical performance of industrial hemp fiber reinforced polylactide biocomposites. Composites Part A: Applied Science and Manufacturing, 42(3): 310-319 (2011).

[125] Bledzki, A K, Jaszkiewicz, A Mechanical performance of biocomposites based on PLA and PHBV reinforced with natural fibres – A comparative study to PP. Composites Science and Technology, 70 (12): 1687–1696 (2010).

[126] Huda, M S, Drzal, L T, Mohanty, A K, Misra, M. Chopped glass and recycled newspaper as reinforcement fibers in injection molded poly(lactic acid) (PLA) composites: A comparative study. Composites Science and Technology, 66 (11–12): 1813–1824 (2006).

[127] Bajpai, P K, Singh, I, Madaan, J Development and characterization of PLA-based green composites: A review. Journal of Thermoplastic Composite Materials 2012, 1–30, DOI: 10.1177/0892705712439571, http://jtc.sagepub.com/content/early/2012/03/21/0892705712439571.full.pdf+html (accessed 10 August 2012).

[128] Bledzki, A K, Jaszkiewicz, A, Murr, M, Sperber, V E, Lützkendorf, R, Reußmann, T Processing techniques for natural and wood–fibre composites. In: Properties and per-

formance of natural-fiber composites. Chapter 4, 163–92, Pickering KL (Ed) Cambridge, UK: Woodhead Publishing (2008).

[129] Choi, H-Y and Lee, J- S Effects of surface treatment of ramie fibers in a ramie/poly(lactic acid) composite. Fibers and Polymers, 13(2): 217-223 (2012).

[130] Graupner, N, Herrmann, A S, Müssig, J Natural and man-made cellulose fibre-reinforced poly(lactic acid) (PLA) composites: An overview about mechanical characteristics and application areas. Composites: Part A, 40(6-7): 810–821 (2009).

[131] Oksman, K, Skrifvars, M, Selin, J F. Natural fibres as reinforcement in polylactic acid (PLA) composites. Composites Science and Technology, 63(9):1317–24 (2003).

[132] Avella, M, Bogoeva-Gaceva, G, Bużarovska, A, Errico, M E, Gentile, G, Grozdanov, A. Poly(lactic acid)-based biocomposites reinforced with Kenaf fibers. Journal of Applied Polymer Science, 108(6): 3542–3551 (2008).

[133] Bax, B and Mussig, J. Impact and tensile properties of PLA/cordenka and PLA/flax composites. Composites Science and Technology, 68(7-8): 1601–1607 (2008).

[134] Nakamura, R, Goda, K, Noda, J, Ohgi, J High temperature tensile properties and deep drawing of fully green composites. Express Polymer Letters; 3(1): 19–24 (2009).

[135] Matkó Sz., Toldy, A, Keszei, S, Anna, P, Bertalan, Gy, Marosi, Gy. Flame retardancy of biodegradable polymers and bio-composites. Polymer Degradation and Stability, 88(1): 138–145 (2005).

[136] Bourbigot, S, Fontaine, G, Duquesne, S, Delobe, R. PLA nanocomposites: quantification of clay nanodispersion and reaction to fire. International Journal of Nanotechnology, 5(6-8): 683-692 (2008).

[137] Shumao, L, Jie, R, Hua, Y, Tao, Y, Weizhong, Y. Influence of ammonium polyphosphate on the flame retardancy and mechanical properties of ramie fiber-reinforced poly(lactic acid) biocomposites Polymer International, 59(2): 242–248 (2010).

[138] Jangm, J Y, Jeong, T K , Oh, H J, Youn, J R, Song, Y S. Thermal stability and flammability of coconut fiber reinforced poly(lactic acid) composites. Composites Part B: Engineering , 43(5): 2434–2438 (2012).

[139] Jiménez, A and Ruseckaite, R A. Nano-Biocomposites for Food Packaging In: Environmental Silicate Nano-Biocomposites Green Energy and Technology, Chapter 15, 393-408, L. Avérous and E. Pollet (Eds) Springer-Verlag, London (2012).

[140] Sanchez-Garcia, M D and Lagaron, J M. On the use of plant cellulose nanowhiskers to enhance the barrier properties of polylactic acid. Cellulose, 17(5): 987-1004 (2010).

[141] Sanchez-Garcia, M D and Lagaron, J M Novel clay-based nanobiocomposites of biopolyesters with synergistic barrier to UV light, gas and vapour. Journal of Applied Polymer Science, 118(1): 188-199 (2010).

[142] Petersson, L and Oksman, K. Biopolymer based nanocomposites: Comparing layered silicates and microcrystalline cellulose as nanoreinforcement. Composites Science and Technology, 66 (13): 2187-2196 (2006).

[143] Lasprilla, A J R, Martinez, G A R, Lunelli, B H, Jardini, A L, Maciel Filho, R. Poly-lactic acid synthesis for application in biomedical devices - A review. Biotechnology Advances, 30(1): 321-328 (2012).

[144] Chen, L, Yang, J, Wang, K, Chen, F, Fu, Q. Largely improved tensile extensibility of poly(L-lactic acid) by adding poly (ε-caprolactone). Polymer International, 59(8): 1154-1161 (2010).

[145] Walker, P A, Aroom, K R, Jimenez, F, Shah, S K, Harting, M T, Gill, B S, Cox Jr, C S. Advances in progenitor cell therapy using scaffolding constructs for central nervous system injury Stem Cell Reviews and Reports, 5(3): 283–300 (2009).

[146] Zhou, W Y, Lee, S H, Wang, M, Cheung, W L, Ip, W Y. Selective laser sintering of porous tissue engineering scaffolds from poly (L-lactide)/carbonated hydroxyapatite nanocomposite microspheres Journal of Materials Science-Materials in Medicine, 19: 2535-2540 (2008).

[147] Hong, Z K, Qiu, XY, Sun, J R, Deng, M X, Chen, X S, Jing, X B. Grafting polymerization of l-lactide on the surface of hydroxyapatite nano-crystals. Polymer, 45(19): 6699-6706 (2004).

[148] Ray, S S and Okamoto, M. Biodegradable Polylactide and Its Nanocomposites: Opening a new dimension for plastics and composites. Macromolecular Rapid Communications, 24(14): 815-840 (2003).

[149] Paul, M-A, Delcourt, C, Alexandre, M, Degée, Ph., Monteverde, F, Dubois, Ph. Poly-lactide/montmorillonite nanocomposites: study of thehydrolyticdegradation Polymer Degradation and Stability, 87(3): 535–542 (2005).

[150] Kumar, R, Yakubu, M K, Anandjiwala, R D. Biodegradation of flax fiber reinforced poly lactic acid. Polymer Letters, 4(7): 423–430 (2010).

[151] Iovino, R, Zullo, R, Rao, M A, Cassar, L, Gianfreda, L. Biodegradation of poly(lactic acid)/starch/coir biocomposites under controlled composting conditions. Polymer Degradation and Stability, 93(1): 147–157 (2008).

[152] Jandas, P J, Mohanty, S, Nayak, S K, Srivastava, H. Effect of surface treatments of banana fiber on mechanical, thermal, and biodegradability properties of PLA/banana fiber biocomposites. Polymer Composites, 32(11): 1689–1700 (2011).

[153] Jandas, P J, Mohanty S, Nayak S K. Renewable Resource-Based Biocomposites of various surface treated banana fiber and poly lactic acid: Characterization and Biodegradability. Journal of Polymers and the Environment, 20(2): 583-595 (2012).

[154] Singh, S, Mohanty, A K, Sugie, T, Takai, Y, Hamada, H. Renewable resource based biocomposites from natural fiber and polyhydroxybutyrate-co-valerate (PHBV) bioplastic. Composites: Part A, 39(5): 875–886 (2008).

[155] Jiang, L, Huang, J, Qian, J, Chen, F, Zhang, J, Wolcott, M P, Zhu, Y. Study of Poly(3-hydroxybutyrate-co-3-hydroxyvalerate) (PHBV)/bamboo pulp fiber composites: Effects of nucleation agent and compatibilizer. Journal of Polymers and the Environment, 16(2): 83-93 (2008).

[156] Nagarajan, V, Misra, M, Mohanty, A K New engineered biocomposites from poly(3-hydroxybutyrate-co-3-hydroxyvalerate) (PHBV)/poly(butylene adipate-co-terephthalate) (PBAT) blends and switchgrass: Fabrication and performance evaluation. Industrial Crops and Products, 42: 461–468 (2013)

[157] Ahankari, S S, Mohanty, A K, Misra, M. Mechanical behaviour of agro-residue reinforced poly(3-hydroxybutyrate-co-3-hydroxyvalerate), (PHBV) green composites: A comparison with traditional polypropylene composites. Composites Science and Technology, 71(5): 653–657 (2011).

[158] Yu, H-Y, Qin, Z-Y, Liu, Y-N, Chen, L, Liu, N, Zhou, Z. Simultaneous improvement of mechanical properties and thermal stability of bacterial polyester by cellulose nanocrystals Carbohydrate Polymers, 89(3): 971–978 (2012).

[159] Ten, E, Bahr, D F, Li, B, Jiang, L, Wolcott, M P. Effects of Cellulose Nanowhiskers on mechanical, dielectric and rheological properties of poly(3-hydroxybutyrate-co-3-hydroxyvalerate)/cellulose nanowhisker. Composites Industrial & Engineering Chemical Research, 51 (7): 2941–2951 (2012).

[160] Sanchez-Garcia, M D, Lopez-Rubio, A, Lagaron, J M. Natural micro and nanobiocomposites with enhanced barrier properties and novel functionalities for food biopackaging applications. Trends in Food Science & Technology, 21(11): 528-536 (2010).

[161] Chen, G X, Hao, G J, Guo, T Y, Song, M D, Zhang, B H. Structure and mechanical properties of poly(3-hydroxybutyrate-co-3-hydroxyvalerate) (PHBV)/clay nanocomposites. Journal of Materials Science Letters, 21(20):1587–1589 (2002).

[162] Nguyen L. T. H., Chen S., Elumalai N. K., Prabhakaran M. P., Zong Y., Vijila C., Allakhverdiev S. I., Ramakrishna S. Biological, chemical, and electronic applications of nanofibers. Macromolecular Materials and Engineering, 297(11): 1035–1123 (2012).

[163] Hazer, D B, Kılıçay, E, Hazer, B. Poly(3-hydroxyalkanoate)s: Diversification and biomedical applications. A state of the art review. Materials Science and Engineering C, 32(4): 637–647 (2012).

[164] Cool, S M, Kenny, B, Wu, A. Nurcombe V., Trau M., Cassady A. I., Grøndahl L. Poly(3-hydroxybutyrate-co-3-hydroxyvalerate) composite biomaterials for bone tissue regeneration: In vitro performance assessed by osteoblast proliferation, osteoclast

adhesion and resorption and macrophage proinflammatory response. Journal of Biomedical Materials Research Part A, 82: 599-610 (2007).

[165] Batista, K C, Silva, D A K, Coelho, L A F, Pezzin, S H, Pezzin, A P T. Soil biodegradation of PHBV/peach palm particles biocomposites Journal of Polymers and the Environment, 18(3): 346-354 (2010).

[166] Petersson, S, Jayaraman, K, Bhattacharyya, D. Forming performance and biodegradability of wood-fibre BiopolTM composites. Composites Part A, 33: 1123-1134 (2002).

[167] Avella, M, La Rota, G, Martuscelli, E, Raimo, M, Sadocco, P, Elegir, G, Riva, R. Poly (3-hydroxybutyrate-co-3-hydroxyvalerate) and wheat straw fibre composites: thermal, mechanical properties and biodegradation behaviour. Journal of Materials Science, 35(4): 829-836 (2000).

Permissions

The contributors of this book come from diverse backgrounds, making this book a truly international effort. This book will bring forth new frontiers with its revolutionizing research information and detailed analysis of the nascent developments around the world.

We would like to thank Rolando Chamy and Francisca Rosenkranz, for lending their expertise to make the book truly unique. They have played a crucial role in the development of this book. Without their invaluable contribution this book wouldn't have been possible. They have made vital efforts to compile up to date information on the varied aspects of this subject to make this book a valuable addition to the collection of many professionals and students.

This book was conceptualized with the vision of imparting up-to-date information and advanced data in this field. To ensure the same, a matchless editorial board was set up. Every individual on the board went through rigorous rounds of assessment to prove their worth. After which they invested a large part of their time researching and compiling the most relevant data for our readers. Conferences and sessions were held from time to time between the editorial board and the contributing authors to present the data in the most comprehensible form. The editorial team has worked tirelessly to provide valuable and valid information to help people across the globe.

Every chapter published in this book has been scrutinized by our experts. Their significance has been extensively debated. The topics covered herein carry significant findings which will fuel the growth of the discipline. They may even be implemented as practical applications or may be referred to as a beginning point for another development. Chapters in this book were first published by InTech; hereby published with permission under the Creative Commons Attribution License or equivalent.

The editorial board has been involved in producing this book since its inception. They have spent rigorous hours researching and exploring the diverse topics which have resulted in the successful publishing of this book. They have passed on their knowledge of decades through this book. To expedite this challenging task, the publisher supported the team at every step. A small team of assistant editors was also appointed to further simplify the editing procedure and attain best results for the readers.

Our editorial team has been hand-picked from every corner of the world. Their multi-ethnicity adds dynamic inputs to the discussions which result in innovative outcomes. These outcomes are then further discussed with the researchers and contributors who give their valuable feedback and opinion regarding the same. The feedback is then collaborated with the researches and they are edited in a comprehensive manner to aid the understanding of the subject.

Apart from the editorial board, the designing team has also invested a significant amount of their time in understanding the subject and creating the most relevant covers. They scrutinized every image to scout for the most suitable representation of the subject and create an appropriate cover for the book.

The publishing team has been involved in this book since its early stages. They were actively engaged in every process, be it collecting the data, connecting with the contributors or procuring relevant information. The team has been an ardent support to the editorial, designing and production team. Their endless efforts to recruit the best for this project, has resulted in the accomplishment of this book. They are a veteran in the field of academics and their pool of knowledge is as vast as their experience in printing. Their expertise and guidance has proved useful at every step. Their uncompromising quality standards have made this book an exceptional effort. Their encouragement from time to time has been an inspiration for everyone.

The publisher and the editorial board hope that this book will prove to be a valuable piece of knowledge for researchers, students, practitioners and scholars across the globe.

List of Contributors

Ganapati D. Yadav and Jyoti B. Sontakke
Department of Chemical Engineering, Institute of Chemical Technology, Matunga, Mumbai, India

S. Sandu and E. Hallerman
Department of Fish and Wildlife Conservation, Virginia Polytechnic Institute and State University, Blacksburg, VA, USA

S. González, E. Pellicer, S. Suriñach and M.D. Baró
Departament de Física, Facultat de Ciències, Universitat Autònoma de Barcelona, Barcelona, Spain

J. Sort
Institució Catalana de Recerca i Estudis Avançats (ICREA) and Departament de Física, Facultat de Ciències, Universitat Autònoma de Barcelona, Barcelona, Spain

Iheoma M. Adekunle
Environmental Remediation Research Group, Department of Chemical Sciences (Chemistry), Federal University Otuoke, Bayelsa State, Nigeria

Augustine O. O. Igbuku and Philip D. Shekwolo
Restoration of Ogoniland Project Team, Shell Petroleum Development Company, Port Harcourt, Nigeria

Oke Oguns
Remediation Team, Shell Petroleum Development Company, Port Harcourt, Nigeria

P. Guerra, J. Amacosta, T. Poznyak and S. Siles
Superior School of Chemical Engineering, National Polytechnic Institute (ESIQIE-IPN), Mexico

A. García
Instituto Tecnológico de Estudios Superiores de Monterrey, Campus Guadalajara (ITESM), Mexico

I. Chairez
Professional Interdisciplinary Unit of Biotechnology, National Polytechnic Institute (UPIBI-IPN), Mexico

Derval dos Santos Rosa
Universidade Federal do ABC, SP, Brazil

Denise Maria Lenz
Universidade Luterana do Brasil, RS, Brazil